STUDIES IN THE CHRONOLOGY AND REGIONAL STYLE OF OLD BABYLONIAN CYLINDER SEALS

Bibliotheca Mesopotamica

Primary sources and interpretive analyses for the study of Mesopotamian civilization and its influences from late prehistory to the end of the cuneiform tradition

Published Under the Auspices of
IIMAS
The International Institute for Mesopotamian Area Studies

BIBLIOTHECA MESOPOTAMICA

Volume 23

STUDIES IN THE CHRONOLOGY AND REGIONAL STYLE OF OLD BABYLONIAN CYLINDER SEALS

Lamia al-Gailani Werr

UNDENA PUBLICATIONS

Malibu 1988

ABSTRACT

Time spans for cylindrical seals used during the Old Babylonian period, c. 2000-1600 B.C., have not been clearly established. This work uses dated tablets from sites with stratified levels—Tell Harmal, Tell Dhiba'i, Larsa, Ur, and Sippar—to identify chronological and regional styles. A catalogue lists (A) seal impressions by museum number, catalogue number, and date, and (B) seal impressions in iconographic categories, by site. A pictorial index to the iconography is furnished by 44 plates of drawings, keyed to the catalogue, and 58 pages of photographs arranged according to tablet and envelope placement.

Published with the assistance of AICF—The Ambassador International Cultural Foundation.

Library of Congress Card Catalog Number: 88-50617

ISBN 0-89003-172-X (paper)
0-89003-173-8 (cloth)

Undena Publications, P.O. Box 97, Malibu, California, U.S.A. 90265

To the memory of

DR. FARAJ BASMACHI

Acknowledgments

This study was originally submitted for a Ph.D. thesis at the Institute of Archaeology, University of London, in June 1977. Professor Edith Porada, who read the paper, suggested its publication, and under her guidance the work was subsequently revised with plates and appendices added.

It is not easy to recount here the many people to whom I am indebted at both stages of my work and whose help has been invaluable. My appreciation and gratitude, however, go foremost to Barbara Parker (now Lady Mallowan), who most generously gave much of her time in directing the course of my research and unfailingly offered much needed encouragement and criticism. I owe similar thanks to Professor David Oates, who advised me in the final stages of the research.

It is Professor Porada who ultimately stands out as the moving force which brought this publication to fruition. To her go my heartfelt thanks for the many invaluable suggestions and criticism she made during her reading of the text.

I wish to thank the Antiquities Department of Iraq and the Iraq Museum (particularly the assistance given by Mrs. Rasmiya Rashid), the Western Asiatic Department of the British Museum, and the Musée du Louvre, for all the facilities provided to me to examine their collections of seal impressions.

I also wish to thank IIMAS–The International Institute for Mesopotamian Area Studies for its financial assistance concerning the photographic documentation.

My gratitude goes to the library staff of the Institute of Archaeology which has been of constant help to me. The same may be said of the Photography Department (Peter Durrel and his assistant, Stewart Laidlaw), and Mudhir el-Khalidi of the Antiquities Department of Iraq, for their technical help in producing the photographic illustrations. Not least I am grateful to Niam el-Hashimi for her helpful suggestions in reading through the manuscript and correcting the English.

TABLE OF CONTENTS

INTRODUCTION

The Old Babylonian period, including the dynasties of Isin and Larsa lasted approximately 400 years, c. 2000-1600 B.C. The style of the seals produced during this time is easily distinguishable, but the criteria for dating them within this period, such as R. M. Boehmer has produced for Akkadian seals, have not yet been established. The purpose of this work is to use certain groups of tablets, which I was fortunate to be permitted to study by courtesy of the Iraq Directorate General of Antiquities, to make some contribution toward the chronology of seals within this period and to attempt a preliminary identification of the products of some of the local seal cutters.

H. Frankfort[1] was misled by the seals he found in the excavation at Tell Asmar, for he believed that inferior works exhibiting an intermediate phase between the seals of Ur III and the full Old Babylonian style were characteristic of the Isin-Larsa period. This was disproved by Edith Porada when she published impressions on tablets from Larsa, dating to Sin-iddinam (c. 1849 B.C.),[2] which showed that the Old Babylonian style was already fully developed by that date, and, in fact, that the finest work was produced before the reign of Hammurabi.

W. Nagel[3] published a very useful list of Old Babylonian seals dated by inscriptions which referred to known rulers and demonstrated that the change of style from that of the Third Dynasty of Ur to that of the Larsa Dynasty began early in that dynasty. An outstanding example of a fine seal of early Larsa in the Iraq Museum refers to King Abisare (c. 1905).[4]

The glyptic art of the latter part of the First Dynasty of Babylon from Ammiditana onward is characterised by much use of the drill, as recognised by H. Frankfort and A. Moortgat.[5] This evidence comes principally from Sippar. In view of the widespread use of the drill after the end of the Old Babylonian period, however, it seems unlikely that the style of Sippar differed greatly from that of other contemporary cities.

A study of seal impressions found on dated tablets that come from archives or from sites with stratified levels is the best means of clarifying both chronology and regional styles. The material available to me were groups of tablets from Tell Harmal, Tell Dhiba'i, Larsa and Ur, mostly dating to the Isin-Larsa period. The material dating to Hammurabi and later is

[1] See *Diyala.*
[2] *Porada,* pp. 155-162.
[3] Nagel, *AfO* 18, pp. 321-327.
[4] Porada and Basmachi, *Sumer* 7, pp. 66-68.
[5] *Cyl. Seals,* p. 180; *VR,* p. 44.

largely restricted to Sippar and its suburb Sippar-Amnanum (Tell ed-Der), the latter excavated by the Iraq Antiquities Department.[6]

I am aware that in my study the lack of material from other sites in Babylonia dating to Hammurabi and later leaves several questions unanswered particularly the geographical distribution of the drilled style mentioned above. The drawings in this study are intended to provide a pictorial index to the iconography of the Old Babylonian sealings. Certain types, like the bull-man, are represented in the drawings in a more or less uniform way without indication of the specific stylistic characteristics of the respective sealing. Such characteristics can only be seen in the photographic reproductions. The seal inscriptions are only sketched in to complete the design and may not be entirely accurate.

On Plates I-XLIV the drawings of the impressions are arranged in conjunction with the iconographical observations in the text. On Illustration pages 1-57 the photographs are arranged according to their placement on the tablets and envelopes and in the Catalogue.

The tablet number under each impression on Plates I-XLIV (in parentheses) and under each photograph on Illustration pages 1-57 directs the reader to the Catalogue description where the museum number and other information are given.

[6]T. Baqir, M. Mustafa, *Sumer* 1, pp. 37-54.

CHAPTER 1

THE DIYALA REGION

1.1. Tell Harmal

There is a group of mounds just east of Baghdad which, judging by their surface remains, flourished during the early Old Babylonian period. Tell Harmal, Tell Dhiba'i and Tell Mohamed[1] are the most prominent among them.

Tell Harmal, a small mound about 10 kilometers south-east of Baghdad, is 150 meters in diameter and 4 meters high. It was excavated by the Directorate General of Antiquities in a salvage operation in connection with a housing scheme. The finds were so spectacular that more than nine seasons of excavation took place between 1945 and early 1962.[2] The mound is rectangular in shape and surrounded by a fortification wall with one (gateway) entrance from the north-east. One large temple, guarded at its entrance by two terracotta statues of crouching lions, was unearthed. Two smaller shrines were also excavated in addition to a number of public buildings, shops and private houses. The site yielded several building levels; those that are important for the present study are Levels II, III and IV which brought to light over two thousand tablets. A large number of these tablets have already been studied by Taha Baqir who published the date formulae,[3] mathematical texts,[4] and the laws of Eshnunna.[5] A. Goetze published the law tablet, which he attributed to King Bilalama of Eshnannu (c. 1990),[6] and fifty letters from a later period.[7]

Various aspects of the tablets have also been the subject of the researches of several students. Amer Suleiman studied the land tenure[8] and a number of the adoption cases. Ridha Al-Hashimi worked on the purchase contracts of fields, houses and slaves.[9] The formulae that occur on some of the tablets included in his thesis are published in *Sumer*.[10] Abdul-Karim Abdullah Ahmed dealt with the loan contracts.[11] The last in this trend is Maria Ellis who included seventy-four administrative tablets from Harmal in her Ph.D. thesis.[12]

[1] Baqir, *Sumer* 2/2, p. 22.
[2] *Ibid.*, p. 22 ff. for a preliminary report on the excavation.
[3] Baqir, *Sumer* 5/1, pp. 34-86; 5/2, pp. 136-143.
[4] Baqir, *Sumer* 6/1, pp. 39-54, and 6/2, pp. 130-149.
[5] Baqir, *Sumer* 4, pp. 63-102.
[6] Goetze, *Sumer* 4/2, pp. 63-102.
[7] Goetze, *Sumer* 14, pp. 3-78.
[8] See *Suleiman*.
[9] See *Al-Hashimi*.
[10] Al-Hashimi, *Sumer* 28 (1972), pp. 29-33.
[11] See *Karim*.
[12] Ellis, "Taxation and Land Revenues in the Old Babylonian Period." Ph.D. dissertation, Yale University, 1969.

It is from this abundant research that I have drawn my material for the study of the glyptic art of Tell Harmal. The majority of the tablets come from Levels II, III and IV. The date formulae from Level IV belong to a number of local rulers;[13] the only historically known king is Sin-abushu, who seems to have gained control over a great part of the Diyala region.[14] According to the stratification at Tell Harmal he should be dated between the reigns of Sumuabum and Sumulael, kings of Babylon (ca. 1894-1845).[15]

The date formulae from Level III belong to Hammidashur, Ibalpiel I and Naram-Sin of Eshnunna, while Level II is assigned to Dadusha and King Ibalpiel II. Level II was destroyed by fire, which was probably caused by Hammurabi's conquest of the Kingdom of Eshnunna in his 31st year.[16]

The history of Harmal, as ascertained from the archives, depicts an agricultural community, under the control of Eshnunna after Level IV, i.e., during Levels III and II, until the conquest by Hammurabi. The ancient name of Tell Harmal, "Shaduppum," [17] is derived from the Sumerian meaning "treasury, accountant's office." This definition is well substantiated by the contents of the tablets. The name Belgasher which is mentioned in several loan contracts has been identified as the main god of the town. His name occurs on a seal impression, as "Belgasher King of Shaduppum." [18]

The majority of the seal impressions on the Harmal tablets were rolled on the envelopes of business documents (sales, loans) and on a few letters, particularly those from Levels IV and III. In Level II a change in the sealing practice seems to have occurred as the majority of the impressions were rolled on smaller tablets. The abundance of impressions and the varied repertoire depicted in the glyptic art of the Old Babylonian period recommend dividing these impressions according to their subject matter.

1.1.1. Contest Frieze of Fighting Figures

Various types of the fighting frieze, typical of the Old Babylonian period are represented in the impressions at Harmal.

Pl. I: 1 is a fragment of a scene which shows two pairs of antithetical contestants as are those in Pl. I: 2. The fragmentary impression belongs to an adoption document which has the only date formula belonging to Harmal, as opposed to Eshnunna and other Diyala

[13] See notes 3, 9, 10, 11; for a historical survey of the region see Frankfort, *et al.*, 1940, chapter 5.
[14] See *Raschid.* The origin of this archive is definitely in the Diyala region although its exact locality is unknown. A number of the date formulae on the tablets could be correlated with those from Harmal. It can be determined from the evidence in these documents, that Sin-abushu was a contemporary of Ibiq-Adad I of Eshnunna or of his son Abdi-Erah. In addition to Harmal and to his own capital, as yet unidentified, he was in control of Ishchali for some time and of a city called Shulgi-nanna (for the latter's location, see Goetze, *JCS* 4, p. 95).
[15] Al-Hashimi, *Sumer* 28, p. 33.
[16] *Ibid.*, p. 30; Hammidashur date formulae also appear in Level IV. No seal impressions are associated with them.
[17] Simmons, *JCS* 13, p. 72.
[18] *Ibid.*, No. 53, p. 73.

sites.[19] Pl. I: 2 presents in its two pairs of contestants a continuation of the favoured compositional scheme of the Agade period. However, in this Old Babylonian impression, the two pairs are identical, each consisting of a bull-man fighting a human-headed bull. This scheme differs from that of the Agade period where the bull-man never has a bull with human or animal head for an adversary.[20] The filler motifs are characteristic in their association with this class of seals[21] and were popular during the early part of the Old Babylonian period.

The fighting frieze depicting more than two pairs of contestants is more usual. Composition of three pairs is seen in Pl. I: 3, 4, 5 and 6. Three participants in a single combat are depicted on Pl. I: 7 and 8.

Although it is agreed that the frieze of fighting figures demonstrates a continuous tradition from the Agade glyptic, and even from that of the Early Dynastic period,[22] the Old Babylonian seal cutter contributed a distinctive composition with his introduction of new motifs such as the lion-griffin.

The athletic struggles of the Agade seals are developed into a new type of frieze in examples such as Pl. I: 6 in which figures of the nude bearded hero with three pairs of curls are represented full face as well as in profile. In Pl. I: 6 three such figures are shown: the first grasps a rampant lion from behind and raises one leg to press it against the animal's back; the second holds a lion in an inverted position; and the third, with head in profile, raises up a lion, probably to hold it above his head in a gesture of triumph. The type of hero shown here can be compared to sealings from other sites such as *Sippar*, Pl. XXXIII: 7 and 8.

Two types of heroes are represented in Old Babylonian contest friezes: the nude bearded hero with three pairs of curls (described above) and a new figure not known from earlier periods. This figure is nude, has a hairstyle which looks like a helmet and a pointed beard. One of its rare representations is found in the Harmal sealing, Pl. I: 4, where the hero raises a lion by the hindleg. A similar figure appears as the victim of a lion-griffin in the same sealing. In contrast to the heroic version of the figure, its depiction as a victim of the lion-griffin or lion is very common in Old Babylonian contest scenes.

A most unusual seal impression is Pl. I: 9, dated to King Sin-abushu. Two crossed lion-griffins have their heads turned back with open jaws in readiness to attack a lion, carried over the head of a kneeling nude hero. As far as I know, this is the only example of such a subject from Mesopotamia in the Old Babylonian Period.[23] Unusual in Mesopotamian glyptic art is the overlapping of figures; here the kneeling hero appears behind the two crossed lion-griffins.

[19] *Karim* No. 40; "mu u-su-sa-e, dingir Be-el- ga-še-er, in-na-du-u," because of the mention of Harmal's main god, Taha Baqir suggests that it is a reference to the building of the main temple at Harmal (See *Baqir*).
[20] See for example, *Corpus*, p. 43, Pls. XXIV and LI.
[21] *Ibid.*, Pl. LI.
[22] Seals of earlier periods did survive and were sometimes even used. Therefore the Old Babylonian seal cutter was familiar with them and could have imitated such seals.
[23] *VR*, 513, an Old Assyrian seal from Assur, shows two crossed lions with their heads turned back. On another seal of peripheral style, two goats are in a similar position: *Yale*, 1173. There is an Agade seal in the British Museum illustrating a mythological scene which has two crossed lions with heads turned backward, *Boehmer*, 276.

The kneeling hero throwing a lion over his head appears on contest friezes of the Old Babylonian period.[24] He appears as a secondary motif on a worshipping scene from Harmal, Pl. V: 3.[25] A suppliant goddess facing the composition Pl. I: 9; a lizard, drill holes, and an unidentified creature (fish, lion-fish?) seems to be swallowing a fly. The second group in this impression shows a human-bearded bull carved over an erased inscription and crossed with a lion; the lion, in turn, attacks a standing goat. The subject may be derived from the Early Dynastic repertoire, but the style of carving is Old Babylonian.[26]

In several instances, figures of the contest friezes accompany devotional or other scenes in Old Babylonian cylinders. Examples are found in our seal impressions. For example, a pair of contestants accompanies a devotional scene in Pl. I: 10, where the nude bearded hero with three pairs of curls raises a lion by one of its hind legs.[27] A bull-man is in combat with the nude hero in Pl. I: 11, and with a lion in Pl. I: 12. A lion attacks a goat sitting on a knoll in Pl. XII: 1. A rather rare motif is seen in Pl. I: 13, depicting two lions juxtaposed,[28] and a lion-griffin holding a crescent standard.[29]

Certain filling motifs are associated regularly with particular scenes. This may have been due to the fact that they belonged to a certain local tradition or to one manufacturing center. However, these may also have been a matter of chronology; some associated features are found in the early Old Babylonian phase, others in the later phase. For example, the two pairs of contestants of the early part of the Old Babylonian period invariably depict the ball-staff, Pl. I: 1 and 7, the double lion-headed club, Pl. I: 2, and mongoose and crouching dog, Pl. I: 7.[30] In fighting friezes where the figures are more cramped, filling motifs are less frequent.

1.1.2. Worship of an Enthroned Deity

In Levels IV and III at Harmal, the majority of the seal impressions depict a worshipper in the presence of an enthroned king or god, a subject which has already been discussed by many scholars.[31] The seals from Harmal conform to the criteria established in the previous discussion; the deified figure of the "king" has a mop of curly hair indicated by a series of short upright lines, Pl. II: 1, 2 and 3, or a cap with upturned brim, Pl. II: 4 5, 6, 7, 8, 9

[24] *Corpus*, 362, *Bib. Nat.*, 434 and *VR*, 479. Here he seems to be associated with the snake goddess.
[25] Also *VR*, 305.
[26] *Cyl, Seals*, p. 148, suggested that the Old Babylonian seal cutters consistently copied certain ancient seals, "thus the lions never cross with another animal." No. 19 contradicts this statement, although it must be pointed out that the group of animals within which the lion crosses the bull forms a secondary motif. This may account for some of the unusual features of the design.
[27] This seal is similar to a seal impression from Ischali, *Diyala*, 942, most probably the work of the same artist.
[28] This may imply a contest of these two creatures comparable to *Legrain*, 448, where a lion fights a winged griffin.
[29] *Louvre* II, Pl. 79, No. 21 (A. 337), where two demons are holding the standard. *Newell*, 163, shows a winged demon also holding such a standard. Other examples: *Yale* 726 and 946; *Yale* 7000 is an impression from Tell Harmal.
[30] For further examples see *Corpus*, Pl. LI-LII; *Cyl. Seals*, Pl. XXVIg. Van Buren, 1959, No. 94; its terminal consists of a panel of inscription above a dog also on *Iraq* 6, Pl. 2, No. 9. Similar terminals appear with devotional and presentation scenes and show their popularity during the early part of the Old Babylonian period.
[31] *Cyl. Seals*, p. 158; *Corpus*, pp. 37 and 41; Van Buren, *ZA* 50, pp. 92-120.

and 10. He wears a simple long garment, doubtless a piece of material wrapped around the body with a fringe falling down the side with one shoulder covered, and the other bare.[32] The throne is usually a stool covered with flounced material, but occasionally, as in Pl. II: 1 and 3, it is a panelled stool in the shape of a temple facade. Frequently, the throne is placed over a platform decorated with a buttress representing the recessed effect of a temple wall. The "king" holds a cup in his hands as if he were tending it to the worshipper who stands immediately in front of him in most cases. The worshipper is usually bareheaded with hands clasped, wearing a long garment wrapped round one shoulder leaving the second bare. Exceptions are Pl. II: 3 and Pl. III: 3, where the worshipper is being led toward the enthroned "king" by an interceding goddess.

It is unusual for a Mesopotamian seal to have the box of inscription below a presentation scene as in Pl. II: 1.[33] The attitude of the arms of the small figure of the worshipper is also known on seals from Mesopotamia, although it appears frequently on seals from Susa.[34] The seal itself is not Elamite but was probably produced in a peripheral region.

The style of Pl. II: 3 is related to seals from the Diyala region, especially the crescent standard with tassels.[35] Pl. II: 11 is unusual because the enthroned figure is placed on the left whereas, since the end of Agade, it is always placed on the right in Babylonia. However, the Agade practice of placing the enthroned deity on right or left is still to be found on the Old Assyrian seals of Kültepe so that No. 11 may be a seal following that tradition.

1.1.3. Enthroned God

A number of impressions from Levels IV and III illustrate the presentation of a worshipper to an enthroned god who normally sits on a panelled stool wearing a horned crown and flounced garment. The god's identity is difficult to determine, if he carries no distinctive insignia and the smaller motifs in the field are not necessarily related to him. Normally, a goddess appears before the god leading a worshipper by the hand. In Pl. III: 1 and 2, the god is enthroned on a stool covered with a flounced material. The suppliant goddess in Pl. III: 2, wears a necklace with counterweight. The high quality of the engraving of the original seal and the manner of the rendering of the god's garment is reminiscent of a sealing of the Ur III period from Ur.[36] The same envelope has two more impressions of presentation scenes. In Pl. III: 3, the interceding goddess wears a triple necklace and a bull-man holding a standard stands at the end of the scene. In view of the frequent association of this figure with the

[32] This is not always clear in the impression due to careless rolling of the actual seals.

[33] A curious seal, *Louvre* II, Pl. 76, No. 25 (A. 251), has such a box, with a fish-man above it. On No. 22 of the same plate, a late Old Assyrian cylinder, the enthroned figure with the hairstyle described above for the enthroned figure in our Nos. 6, 16a and 42a from Tell Harmal, holds a serrated saw. *Yale* 758 is a seal impression showing two facing suppliant goddesses over an inscription.

[34] *Susa*, Pl. 160, Nos. 1736-38. It is interesting that this gesture does not occur in the cylinders of the Diyala region, although one would expect strong Elamite influence in this material for geographical and political reasons.

[35] *Diyala*, 740 and 771.

[36] *Ur Ex.* 10, Pls. 26-28 passim.

Sun God, it is possible to assume that the enthroned god shown here was meant to represent that god.[37] The filling motif in Pl. III: 4 is a small figure of a Snake Goddess.[38]

Though the impression of Pl. III: 5 is faint and partly damaged, it is clear that the enthroned god rests his feet on a gazelle which identifies him as the god Amurru.[39] The impression of Pl. III: 6 looks like an infant on the lap of the deity. If this is correct it would be the only known Old Babylonian example of a motif common on Agade cylinders.[40]

Inscriptions accompanying the presentation scene generally consist of two columns: Dingir Shamash; Dingir A-A, Pl. III: 7, 8 and 9. No. 8 has a dog beneath the box of inscription which is a popular motif on early Old Babylonian seals.[41]

1.1.4. The God with Ring and Staff

In the previous section the discussion dealt with an enthroned god of no particular identity. A few impressions from levels IV and III, Pl. IV: 1, 2, 3, and 4, show a god holding in his hand an emblem termed the "ring and staff," which are invariably held by a number of major gods. These insignia have been the subject of many theories concerning their meaning and symbolism.[42] On the envelope of Pl. IV: 1, a worshipper offers a kid to the enthroned god who holds a ring and a staff.[43] It is probable that the Sun God was meant to be shown here, as in Pl. IV: 2 of Sin-abushu date, because the deity rests its feet on a reclining human-headed bull, a creature associated with the Sun God.[44] The original seal of Pl. IV: 3 had a grooved border made with the bulge, or bulges, circling the metal caps. These are often found on seals of this period and of the preceding Third Dynasty of Ur.[45]

A god in an ascending posture holding a ring receives a libation from a worshipper in Pl. IV: 5. The god's posture is similar to that of the Sun God, who usually places his foot on a miniature mountain.[46] The god with the ring in No. 5 is therefore likely to be Shamash.

[37] *Sippar,* p. 104, seal XVI.

[38] *Cyl. Seals,* Pl. XXXa; Buchanan, *Iraq* 33, pp. 1-18, for a discussion of this goddess.

[39] *Louvre* II, Pl. 80, No. 6 (A. 362) shows a god holding a saw and stepping on a gazelle, probably a confusion with the god Amurru. *Cyl. Seals,* p. 161. Kupper, 1961, p. 27.

[40] For examples see *Boehmer,* Pl. XLVII, 555-560.

[41] Examples are: *Corpus,* 347, 411; *Louvre,* II, Pl. 78 No. 13; *Danish Ntl. Museum,* 50; *Ashmolean,* 447; *VR*, 377; *Newell,* 173; and *Yale* 980. It also occurs in our No. 30 from Level IV, and in No. 144 of the archive of Nurshamash. The motif is derived from Agade seals such as *Boehmer,* 228 and 274; *Corpus,* 144; *Newell,* 100.

[42] Frankfort suggested that they are a measuring instrument and line: *Cyl. Seals,* p. 179; while Van Büren, 1945, p. 144 rejected that suggestion and stated that "the rod is probably a symbol of wisdom and the ring, a symbol of power." E. Sollberger and J. Kupper have lately come to the conclusion that the ring is the "*ṣerretu*," the symbol of the reins of power, *IRSA*, p. 196, note. 1.

[43] The date formula of this document mentions the god Belgasher, the main deity at Harmal.

[44] *Corpus,* 391; *Louvre* II, Pl. 80, No. 23 (A. 350).

[45] For an example in an extant cylinder see *Louvre* II, Pl. 74, No. 18 (A. 191).

[46] *Cyl. Seals,* p. 171.

This identification is supported by the fact that Shamash is the recipient of a libation in an impression from the archive of Nurshamash, Pl. V: 9.[47]

The impression of Pl. IV: 4 from Level II at Tell Harmal shows an enthroned god holding a ring and staff.[48] He rests his feet on a serpent dragon, whose body extends underneath the throne. The god receives a bearded god who leads a worshipper by the hand.

The manner of execution of the seal varies from the contemporary known cylinders and sealings of the Diyala region as both the presentation scene and the enthroned deity had disappeared from the seals at Harmal by the time of Level II. Furthermore, the crown of the god seen in profile is unlike any example from Tell Harmal or, in fact, any seal excavated from Mesopotamia. In Babylonian reliefs and seals the face of the deity is usually rendered in profile while the crown is always rendered from the front.[49] The practice survived until the time of Hammurabi. In his stela, however, the crown of the enthroned Shamash is depicted in profile like the rest of the figure. This has led Moortgat[50] to attribute that part of the Mari paintings, where the only other example of such a crown occurs, to the time of King Zimrilim after his return to Mari, i.e., approximately contemporary with the code of Hammurabi. However, the Harmal impression is earlier than both the paintings and the stela as will be explained later. If this seal is of north Mesopotamian-north Syrian manufacture, the rendering of the crown in profile was an innovation of that artistic region.

The crown of the interceding god is also in profile but here, unlike the several pairs of horns in the crown of the enthroned god, only one pair of horns is attached to a round cap. Again a parallel to such a crown is to be found in the investiture scene at Mari.[51] The worshipper, doubtless a king, wears a high hat rounded in front and rectangular in the back. It is a headdress which closely resembles one seen on a cylinder seal of unknown origin at the Bibliotheque Nationale.[52] The garment of the principal personage on this cylinder, again a king, is decorated with horizontal borders, a decoration for which no precise parallel can be cited at present. The gesture of the king's hands is similar to that of the king in the Harmal impression and of king Zimrilim in the investiture panel at Mari. In the sealings of Mukannishum, a high official of Zimrilim, the king wears a robe with double borders, similar to that of the king in the Harmal example.[53] Similar borders also appear in another cylinder, where the worshipper's cap seems to resemble that of the two examples cited here.[54] The worshipper offers a kid to a god, who is probably Shamash, holding a ring and standing upon two human-

[47] Another example: *Ur Ex* 10, No. 442, dated to king Siniddinam of Larsa.

[48] Ellis, *JCS* 26, p. 138. The envelope belongs to a tablet dealing with a discussion of an estate and dated to the 10th year of Ibalpiel II.

[49] A crown in profile is seen, however, on three Old Babylonian terracotta plaques: from Larsa, Barrelet, 1968, 573; from Nippur, *Nippur I,* Pl. 134, No. 9; and from Khafajeh, A Parrot, 1960, No. 358a, but it should be noted that none of these is similar to No. 76.

[50] Moortgat, 1969, p. 71.

[51] *MAM II,* Pl. XI. It has been suggested that these crowns were made of cloth: Halder, *Orientalia Suecana 1, pp. 51-65.*

[52] *Bib. Nat.,* 428.

[53] *MAM II,* Pl. XLI.

[54] *Cyl. Seals,* Pl. XXVI, k.

headed bulls. In the second compositional group of the same seal an enthroned god who holds ring and rod rests his feet on a serpent dragon.

The serpent dragon was associated with the god Marduk in other cylinders.[55] However, the fact that the seal owner's name in Harmal, Pl. IV: 4, is Tishpak Gamil makes an association with the latter god more likely. Tishpak was the principal god of Eshnunna and the dragon appears to have been his symbol. In a cylinder of the Third Dynasty of Ur which was recut for Kirikiri of Eshnunna,[56] two rearing serpents were secondarily carved as if springing from the god's shoulders. There is also a fragment of an Old Babylonian seal from Khafajeh, showing a throne placed over two dragons.[57]

In addition to the iconographical feature of the serpent dragon which unites the sealing from Harmal Pl. IV: 4 and the cylinder Figure a, both seal owners styled themselves "servant of Shamshi-Adad" in their inscriptions. Thus the seal designs can be dated to the reign of that king of Assyria (c. 1815-1782), and located in the region between Mari and Eshnunna.[58]

1.1.5. The Sun God

Shamash is the most frequently represented deity in the Harmal impressions. His main symbols are the saw or serrated knife and the human-faced bull. He is most often shown standing with his foot on a stylized mountain or a reclining bull. Facing the god is a worshipper wearing a round cap with upturned brim, and a long garment open in front, revealing one bare leg. The worshipper is always represented offering a kid to the god and followed by a suppliant goddess.

A few seals show the Sun God enthroned on a chair with a low back. In Pl. V: 1 he rests his feet on a reclining bull, his throne is rendered with small irregular circles probably representing the god's symbolic mountain. He holds his saw in a vertical position which seems to have been the norm when Shamash is enthroned.[59] A god with bull's ears is depicted in the same impression.[60]

A similar throne, with a high back, is seen in the sealing Pl. I: 10. The actual impression has a strong likeness to a clay sealing, unearthed at the Temple of Shamash as Ishchali.[61] In

[55] *Corpus*, p. 48. The monster was also attributed to Ningizzida by Van Buren, *Iraq* I, pp. 60-89, and to Tishpak by Frankfort in connection with the cylinder of Kirikiri discussed below in note 56.

[56] *Diyala*, 709. The observation that the cylinder was recut from an earlier example of the Third Dynasty of Ur was made by E. Porada on the basis of a study of the original cylinder in which the inscription shows signs of recutting and the serpent necks are obviously added to the shoulders. For a reference to this observation by J. A. Franke see McG. Gibson and R. D. Biggs, 1977, p. 66, note 21.

[57] *Diyala*, 432.

[58] Ellis, *JCS* 26, p. 138 suggests that Tishpak Gamil, the owner of the original cylinder of Harmal 76a, may have come from the Diyala region. Harmal 76a is also discussed by E. Porada, 1979, pp. 259-270, fig. b.

[59] *Cyl. Seals*, Pl. XXVIIa; *Diyala*, 942; *Bib. Nat.*, 146.

[60] Porada suggests a balance of power of the gods represented in the same cylinder seal; see *Corpus*, p. 49.

[61] *Diyala*, 942.

addition to the main scene of presentation, the terminal consists of the nude hero subduing a lion by holding it in an inverted position and placing one leg on its head.[62] The differences appear in the secondary motifs: disc standard, scorpion, vase and drill holes in P. I: 10; in the Ishchali impression only a small figure with a mace is depicted. The distinctive posture of hero and lion, and the type of throne, make it plausible that both original seals were the work of one craftsman, and the secondary motifs were inserted for the purpose of distinguishing individual ownership. The Sun God on Pl. V: 2 is also enthroned on a similar chair.[63] In addition to a standing Shamash, part of what may have been the figure of an enthroned Sun God is visible in the sealing Pl. V: 3. It would be most unusual, however, to have the Sun God portrayed twice in one impression. The rendering of the figures on this seal differs from the rest of the Harmal repertoire e.g., the worshipper with his short beard, the gesture of his hands and the patterned border of his garment and Shamash standing upright wearing a flounced garment.[64] It points to the fact that this seal was an importation, most probably from outside the Diyala region.

A nude attendant stands on a platform and holds a slender curved object behind Shamash in the sealing Pl. V: 4. Perhaps such figures, which frequently appear in association with Shamash[65] were meant to represent statues placed in temples.

The nude priest with pail and sprinkler is another figure which frequently appears in scenes featuring Shamash as in the sealing Pl. V: 5.[66] However, his presence is often the obvious result of the depiction of a deity identified with Nergal in scenes in which Shamash is also shown, Pl. IV: 5.[67]

1.1.6. The Goddess Ishtar

The few impressions representing the goddess Ishtar as the goddess of war come from Levels IV and III. However, Pl. VI: 9 though found in Level II, has a Naram-Sin date formula which places it in the period of this king of Eshnunna. Ishtar is depicted full face, wearing a long pleated skirt and horned crown, her long hair falling down her shoulders in two locks. All her weapons are depicted: arrows in quivers slung over her shoulder, a double lion-headed club which she carries aloft in her right hand and a scimitar, the handle of which she holds

62. For the same motif associated with the Sun God, see *Corpus,* 397; here the nude hero is being helped by a second hero.

63 The tablet is dated by an oath formula to Immerum, a local ruler of Sippar, contemporary of Sumulael of Bablyon.

64 For examples of similar representation of Shamash, see *VR,* 384; *Corpus,* 420.

65 A similar attendant appears behind the Sun God in *Bib. Nat.,* 177; *Danish Ntl. Museum,* 39; *Diyala,* 971, where the author calls the object held by the priest "a sprinkler brush or saw"; *Louvre* II, Pl. 80, No. 9 (A. 353); *Ur Ex* 10, 488; *Corpus,* 419 and 420. Although appearing sometimes with other deities, the figure seems to be most closely associated with Shamash, *Cyl. Seals,* p. 162. Perhaps he is participating in a ritual washing ceremony as suggested by Börker-Klähn, *ZA* 46, pp. 235-240.

66 *Ashmolean,* 494b and 502; *Bib. Nat.,* 167, 212, 229; *VR,* 378; *Gordon,* Pl. III, 16; *Corpus,* 418.

67 *VR,* 376, 377; *Bib. Nat.,* 138. The latter seal belonged to a servant of Sumulael. He is as frequently associated with other Old Babylonian deities as with Shamash, and it is therefore not possible to suggest that he is the same personage as the man with a saw.

with her left hand. She has one foot placed forward upon a crouching lion. In all the examples from Harmal, Pl. VI: 1, 2, 3, 4, 5, 6 and Pl. V: 8, the lion is stylized and simplified leaving only the head recognizable. Unfortunately, the section in front of the goddess is damaged in most of the impressions, which makes it difficult to distinguish the figure of the supplicant. In Pl. VI: 1 a nude attendant with frontal face and three pairs of curls, probably holding a flowing vase, stands before the goddess. [68]

Ishtar is rarely represented in profile as in Pl. VI: 4. Here she holds only a scimitar while a bow is slung over her shoulder in Pl. V: 6. She is also in profile in Pl. VI: 7 of the Nurshamash archive. [69]

At Harmal, where Shamash plays the main role, Ishtar appears together with him if the impressions are complete or nearly so, Pl. V: 6 and 8, Pl. VI: 2, 3 and 8. Her secondary position on some seals and the comparatively small number of representations of the goddess in sealings from Harmal may indicate that Ishtar was not an important deity there. [70]

1.1.7. The God with Scimitar and Other Warring Gods

A god wearing a horned crown and a long garment which covers one shoulder and opens down the front, where there may be two rows of fringe slightly rounded at the bottom, is shown with one leg stepping forward. He has one hand held to his chest, covered by the garment and holds the handle of a scimitar with his other hand. The figure occurs in several sealings from Harmal, Pl. VII: 1, 2 and 3, but is never identified by any special attributes. In Pl. VII: 1, the god faces a priest with pail and sprinkler. [71] In Pl. VII: 2 an attendant with a "sprinkler" follows the god with a scimitar. The sealing Pl. VII: 4 from the Nurshamash archive is so similar to the Harmal example that it is possible to say they may be the work of one seal cutter. [72]

[68] *Diyala,* 940, has a similar composition, but the cylinder is a peripheral product. This classification can be made on the basis of the rendering of the straps of Ishtar's weapons: *Corpus,* 869, 876. Moreover, the peripheral character of that seal is also confirmed by the angular carving of the water flowing from the shoulders of the kilted attendant, and by the figure of the bull-man holding a standard, both popular motifs of the Old Assyrian repertoire.

[69] Other examples of Ishtar in profile are: *Yale* 905; *Corpus,* 539; *VR* 389, 390, 391; *Legrain,* 423; *Bib. Nat.* 169, 234, 238 and 240. On this last seal the double lion club has been distorted; it is detached from the rod which the goddess holds to form an omega-shaped design.

[70] It is also possible that these seals were the product of one centre where the combination of Ishtar and Shamash on one seal was popular.

[71] *Diyala,* 958, shows a similar god facing a worshipper with a priest behind him. *Louvre* I, Pl. 51, No. 4 (D. 35), has the god facing a priest, who also appears on *MAM* II, Fig. 123; *Bib. Nat.,* Pl. XVII, 234. In the cylinder *VR,* 415 the god stands facing right, which is very unusual since all major gods face left in Old Babylonian seals, and has a lion scimitar in front of him and a priest with pail and sprinkler standing behind him. The exceptional arrangement of the figures indicates that the cylinder was made outside one of the principal centres of Babylonia. *Legrain,* 464 has a lion club in the field; the inscription mentions the goddess Kititum, known from the Diyala region, possibly this seal is therefore from that region. *Louvre* II, Pl. 78, No. 27 (A. 315), shows an attendant behind the god. In *Ashmolean,* 494a, the attendant holds a standard and an axe over his shoulder.

[72] In *Bib. Nat.,* 156, the attendant with a sprinkler is also behind the god.

A god with a scimitar and with the unusual feature of bull's ears, represented full face, appears in the sealing Pl. VII: 5.[73] A god with a scimitar appears to be associated with a lion-headed demon who faces the god and holds the head of a human victim in the sealing Pl. VII: 3.[74] The same scene includes a pair of embracing figures. In Pl. IV: 5 the demon holds the head of an animal and faces a god trampling on a human victim who wears a long, pleated skirt. These demons are thought to be demons of disease, and their master, the god with the scimitar, to be the god of the underworld.[75]

A figure clad like a king belongs to the figures which menace and seem to kill humans. Here in Pl. VII: 6, from Level IV, and called by us "royal hero," he is shown in a posture of conquest. In Pl. VII: 7 he carries a multiple mace and tramples on an enemy. An assistant warrior is depicted in some impressions, as in Pl. VII: 7 and 8. The god with scimitar appears in conjunction with the conquering "royal hero" in Pl. VII: 8 and 9.[76] The association of both figures may be of special significance.[77] In Pl. VII: 10, the "royal hero" seems to be associated with the lion-headed demons.[78]

Having discussed the many occurrences of the god with the scimitar on the impressions from Harmal in comparison with other published seals, the identity of the god is uncertain. He

[73] *Louvre* II, Pl. 80, No. 6 (A. 362), and an unpublished seal in the Iraq Museum, IM 23581, show a god with bull's ears in ascending posture wearing a long pleated skirt, stepping on a platform, and carrying a lion scimitar. A worshipper pours a libation in front of the god, followed by a priest with pail and sprinkler. Behind the god is a curious bow-legged figure with an unidentifiable animal over its shoulders. A god with bull's ears in an impression from Sippar dated to the second year of Hammurabi (*Sippar* No. 199.A), who steps on a leonine demon, may connect this god with Nergal, god of the nether world who was master of these demons that brought pestilence and disease. *Corpus*, 395, shows the god holding a lion scimitar and facing the priest with pail and sprinkler with, between them, a double lion club with mace. The lion club is omitted on 396; on 386, the god in a sarcophagus is added. For the connection with Nergal see Porada and Basmachi, *Sumer* 7, pp. 66-68. *MAM* II, Pl. XXXIX, 788 shows a god with scimitar and club facing a worshipper followed by a priest. In *VR*, 383 the god holds an axe over his left shoulder. He also faces a priest.

[74] *Louvre* II, Pl. 78, No. 26 (A. 314). *Corpus*, 381, has the demon brandishing a dagger with one hand, and holding a human victim by its leg, following is a worshipper who faces the god with scimitar, also in *Bib. Nat.*, 445; and *Danish Ntl. Museum*, 81, the demon with a dagger stands facing the god who is on the right side of the impression.

[75] *Cyl. Seals*, p. 167. See above, note 73.

[76] *Corpus*, 382, where a warrior, slinging a bow over his shoulder, assists the conquering king. A worshipper faces the god with scimitar. *Yale*, 901 belongs to the same class of seals.

[77] It may be noted here that of the several impressions on the envelope of Harmal No. 20: a. shows the conquering "king-hero"; b. has a filler motif depicting a lion-headed demon and c. shows the god with the scimitar. This shows the frequency of such themes in the early Old Babylonian period. Other examples of a conquering "king", *Bib. Nat.*, 241 and 242, show a worshipper pouring a libation. In both seals the filler motifs are unusual. They comprise drill holes, a plough, and the foetus-demon, symbol of death. For the last motif see, Porada, *Studies . . . Oppenheim*, 1964, pp. 159-161. In *Legrain*, No. 443, the king is assisted by two male figures.

[78] This is by no means a unique example. *Diyala*, 907, shows a similar composition, identified by Frankfort on this particular seal as the god Nergal, *Cyl. Seals*, p. 167. A combination of several demons on one seal is seen in *Ashmolean*, 532: a lion-headed demon brandishing a dagger, restraining a winged lion-demon who is attacking a bull-man with bird's feet. The latter raises one hand as if asking protection from a god who pours some liquid onto the ground.

could be the god Nergal as identified in his conquering aspect, particularly when holding the double lion-headed club.[79]

In summary it can be stated that three types of gods are represented holding a scimitar—or one god in three different aspects. The first is a god entirely in human form, holding the scimitar with its top set on the ground. The second holds, in addition to the scimitar, a lion club or a mace flanked by two scimitars each topped by a feline head. The third is a god with bull's ears with a scimitar held in the aforementioned posture.

The commonest form represented in the Harmal impressions, and indeed, in any other Old Babylonian seals, is the first type.[80] He is frequently the main figure in a simple ritual scene consisting of a worshipper with a round cap-like headdress, never with the cap with upturned brim or bareheaded. He always has his right hand raised forward in front of his face, while the other is placed at his waist. The long garment with decorated fringed border occasionally reveals one leg. The priest with pail and sprinkler appears immediately facing the god or standing behind the god or the worshipper. Occasionally, the priest stands on a platform. Holding only a sprinkler, he also appears in association with this god.

In contrast to the many worshipping scenes of this period connected with other deities, e.g., Ishtar, Shamash or an enthroned undetermined deity, there is an obvious absence of both the interceding or suppliant goddess from the scenes featuring the god with a scimitar.[81] Certain motifs seem to be associated with the god with the scimitar such as lion-headed demons, and the dog balancing a crook, whose significance is as yet unsolved. The god is also associated with the "royal hero" on a number of seals as seen in the group from the Nurshamash archive and other examples. This same "royal hero" also occurs on seals with a lion-headed demon.

The second version of what may have been the same god shows him with one hand raised in front of his face holding the lion-headed club, a warlike and powerful symbol, and trampling on an enemy. His head, with a long beard, is in profile. He wears a long pleated skirt but no garment covers his upper body. The long pleated skirt, also worn by other major deities, such as Shamash and Ishtar, is open in front to give freedom to a deity's leg when in ascending posture or when trampling an enemy. The god with the double lion club does not participate in worshipping scenes, but is frequently associated with demons who may serve to illustrate his violent nature.

The third figure, the god with bull's ears, always appears with a beard and full face. This is probably due to the iconographic requirement for showing both ears. The god wears the long garment with a fringed border when in a static pose, but wears the pleated skirt when he is

[79] *Cyl. Seals,* p. 167. Porada and Basmachi, *Sumer* 7, pp. 66-68, base the identification on both the iconography and inscription on a seal from Larsa. E. von Weiher, *AOAT* 11, p. 45, suggests that the lion club (Abb. 7), may be the weapon of Nergal, substantiating his argument with its later occurrence on Kassite *kudurrus.* He also suggests that the double lion-headed club may belong to Ninurta, and casts doubt on the validity of Porada's interpretation of the Larsa seal. He argues that in literary texts Nergal is seldom mentioned with the lion which was associated more frequently with Ninurta.

[80] In two exceptional seals the god holds with his folded front hand a lion club: *Corpus,* 395; *MAM* II, 788.

[81] The exception is *Corpus,* 380; where a suppliant goddess is facing him. Nergal is invoked in the inscription. *Yale,* 848 has an inscription separating the god and suppliant goddess.

in an ascending posture.[82] He also seems to participate in ritual scenes like the god with a scimitar.

Although it seems difficult to explain the intricate relationship between the figures of gods described, it is possible they represent particular events and myths known to both the seal cutter and the seal owner.

1.1.8. The Weather God

Adad as a Weather God appears on Old Babylonian seals from as early as the reign of King Sin-abushu, Pl. VIII: 1 and 2. In fact there seems to have been continuity in the god's presence from the Agade period through Ur III and the Old Babylonian periods,[83] to the time when he became a popular figure in the post-Hammurabi era.[84]

There are two ways in which the Weather God is depicted in Old Babylonian glyptic art, doubtless corresponding to two different traditions. One was the tradition of the Agade period, when the symbolic monster of the Weather God was a winged lion-dragon; the other tradition, that of the western peoples of Anatolia and Syria, represented a Weather God standing on a bull.

In Pl. VIII: 1, the god stands on a bull and holds the lightning fork and bull's rein,[85] with a nude female standing behind him.[86] The Weather God also stands on a humped bull in Pl. VIII: 2 and 3;[87] while in Pl. VIII: 4, he is in an ascending posture with a short, kilted god wearing the Syrian top hat associated with the god Amurru facing him.[88] Behind the short, kilted god is a naked figure, also wearing the top hat of Amurru, but holding a crook, Amurru's most distinctive symbol.[89] Amurru and Adad facing each other is a subject encountered several times in Old Babylonian seals.[90]

[82] Unpublished seal in the Iraq Museum, IM 23581.
[83] *Boehmer*, Pl. XXXI; Buchanan, *Iraq* 33, Pl. I.
[84] Porada thought that he appeared in southern Mesopotamia later in the Old Babylonian period under the influence of the glyptic art of the Assyrian merchant colonies in Anatolia, *Corpus*, p. 58.
[85] This is a characteristic posture always to be identified with Adad. See *Cyl. Seals*, p. 162; *Corpus*, p. 58.
[86] A similar iconography can be observed in *Corpus*, Nos. 510 and 511; *Ashmolean*, 521; *VR*, 403-405; on three unpublished seals in the Iraq Museum, IM 13228, 14608 and 21182; and on an Old Assyrian seal, *Bib. Nat.*, 251. In fact she is also seen on seals where his symbols occur: *Corpus*, 493, 503-506; *Louvre* II, Pl. 81, No. 23 (A 394); *Susa*, 1762; *Danish Ntl. Museum*, 56.
[87] For parallels for the humped bull, see *Corpus*, 512; *Diyala*, 909, a rather poorly cut seal.
[88] See Kupper, 1961.
[89] *Ibid.*, p. 35, No. 6.
[90] *Ibid.*, Pl. I, 3; *Susa*, 1766; and an unpublished seal in the Iraq Museum, IM 14213, where Adad stands on a bull and holds a treble lightning fork, facing three columns of inscription. On the other side is Amurru stepping on a gazelle and holding two crooks.

A god, most probably Adad, standing on a lion-dragon is seen in what is visible of the sealing of Harmal, Pl. VIII: 5.[91] The winged dragon associated with Adad is not distinguished by its shape from the lion-headed winged demon who stands on his hind legs in Old Babylonian contest scenes such as those in Pl. XIV: 5 and 7. Only the posture of the dragon, with all four feet on the ground, and the fact that he spits fire differentiates him from the other.[92] However, the fact that the monster rests on its hind legs in scenes in which its demonic character is obvious, may indicate a change in its meaning from the time of the Third Dynasty of Ur.[93]

1.1.9. The Water God

Ea, the Mesopotamian Water God, is portrayed at Harmal by only two seal impressions from the time of Naram-Sin. Pl. VIII: 6 shows part of the god's figure with water flowing from his shoulder. The god sits on a tortoise[94] which is swimming along with some fishes on two streams of water. To the right, one of Ea's nude assistants kneels on one knee.

The second seal, Pl. VIII: 7, depicts Ea holding a vase with water bubbling from it and pouring down on each side of him, to meet with more water flowing from a vase held by a nude assistant, who is kneeling on one knee. The water then flows into round vases only to flow out again, resulting in a stream at the base of the seal. Wavy lines dominate the design of this seal, and even the platform is represented as waves. Ea, with his flowing vase and his nude assistant with three pairs of curls, is seen as early as the Agade period, and into the Neo-Sumerian period. Gudea's seal is the best example.[95] This Harmal impression, Pl. VII: 7, and other examples from the Old Babylonian period are most probably imitations of such earlier seals.[96]

An earlier impression, Pl. VII: 8, of Sin-abushu date, which could be connected with the Water God, shows a kneeling male figure receiving water flowing from a vase held by a tall nude figure with frontal head and three pairs of curls.[97]

[91] An impression from Larsa of Warad-Sin date, *Louvre* II, Pl. 112, No. 10b (A. 485), shows Adad in a similar posture. *Cyl. Seals*, Pl. XXVII,i has Adad enthroned on a fire-spitting winged lion-griffin, a derivative from Agade glyptic. The best example of Adad on a winged monster is *Legrain*, 445, where he also holds the lightning fork and bull's rein.

[92] *Cyl. Seals*, p. 175.

[93] Buchanan, *Iraq* 33, Pl. I d; *Yale*, 679.

[94] The tortoise was associated with Ea in a myth where it saved his temple from assault by the God Ninurta; see Alster, *JCS* 24, pp. 120-129.

[95] *Louvre* I, T. 108 Pl. 10, Fig. 8 and 10. For a similar throne see *Ur Ex.* 10, 398.

[96] The seal of Iluna-kirish from Mari (Amiet, *Syria* 37 p. 215 Fig. 1), shares many motifs of Harmal No. 55a Pl. VIII. A seal from Susa, *Susa,* 1769, depicts mermaids receiving water with round vases from the standing Water God. A good example of an Old Babylonian seal imitating an Agade one comes from Tell Taya, J. Reade, *Iraq* 35, Pl. LXXb.

[97] Two examples where a nude kneeling figure is the recipient of water are *Cyl. Seals*, Pl. XXVIII k and Pl. XXIX l. In both seals the figure is the nude bearded personage with three pairs of curls who frequently functions as the assistant of the Water God although he also appears in association with other gods as in Harmal No. 11a Pl. VI and No. 76b.

In one impression, Pl. VII: 9, a water goddess is depicted standing on a goat fish, with fish in the field. She is probably the goddess Nanshe "mother of the fish,"[98] or Damkina, Ea's wife.[99]

1.1.10. The God with Mace

A god wearing a horned crown and a short tunic which hangs in a point between his legs occurs in a few impressions from Harmal. He frequently holds a mace. Some examples depict him with his hair in a chignon distinctive of gods and wearing a tall cap with a narrow brim rather than the crown of multiple horns seen in Pl. IX: 4 and 5.[100]

He is seen in early impressions from Level IV, Pl. IX: 1 and 2, but is more frequent in Level III, particularly on documents contemporary with Naram-Sin of Eshnunna: Pl. IX: 3, 4, 5, 6, and 7. Facing this god in Pl. IX: 1, 9 is a goddess with a hand gesture now more appropriate for female worshippers than for a goddess.[101]

The impression of Pl. IX: 7, is most uncommon. Behind the god stands a nude goddess wearing the horned crown, and like the nude female, her hands are clasped. Unusual for this figure are the wings which appear to descend from her waist. This is a feature not known before the Mitannian period.[102] Nor is a nude winged female normally found in Old Bablyonian seals, though there are a few examples in terracottas.[103] These figures in terracotta, however, show the wings hanging from the shoulders. They are the only Old Babylonian parallels from Mesopotamia for the Harmal example. A goddess with drooping wings hanging down from her shoulder, with her hands raised upwards, is incised on a pottery jar from Larsa. Her association with the animals decorating the vase may tentatively identify her with Ishtar.[104] Later, at Nuzi, a goddess with drooping wings appears.[105] Her significance is still obscure. Barrelet identifies the spread-winged goddess with the warrior character of Ishtar.[106]

1.1.11. The Figure with Mace

Probably the most common figure in Old Babylonian seals is the one who wears a round cap with upturned brim, a shawl that covers one shoulder forming a tunic which hangs down in a point between the legs. The figure holds a mace by his side with his left hand, while the

[98] M. Civil, *Iraq* 23, p. 154.

[99] Frankfort, *Cyl. Seals,* p. 170. As will be discussed later p. , Damkina was frequently depicted on seals at Sippar.

[100] A seal from Ishchali shows such a headdress more clearly: *Diyala,* 908; also *Louvre* II, Pl. 81, No. 24 (A. 395). This type of cap was popular on the Sippar seals. As yet the god with mace has not been identified with any particular deity; certainly he was not a major one.

[101] *Corpus,* 394; *Diyala,* 913.

[102] For a detailed study see Barrelet, *Syria* 32, pp. 222-260.

[103] *Nippur* I, Pl. 134, No. 6.

[104] *Syria* 32, p. 253, Fig. 19b; Parrot, *RA* 30, p. 180. *Nippur* I, Pl. 134, No. 8 where a phallus on either side of her feet may indicate her fertility character.

[105] Porada, 1947, Pls. XXXVII and LIII.

[106] *Syria* 32, pp. 222-260, and Porada, *AfO* 25, p. 137.

right hand is closed in a fist and appears to hang down the back, although most probably it was meant to hang at the side.

Most of the early impressions from Harmal, from Levels IV and III, depict the figure playing a minor role, normally standing at the end of the scene: Pl. II: 2, 4 and Pl. X: 1.[107] A few examples exist in which a worshipper appears behind the figure as in Pl. X: 2. This is unusual, as worshippers are regularly accompanied by a suppliant goddess. Another example of a worshipper standing behind the figure is seen in Pl. XII: 1.[108] The seal is possibly an heirloom as the complicated scene and the smaller motifs are alien to Level II. As it will be discussed below, in Level II the composition of the seals becomes restricted to the figure with a mace and a suppliant goddess.

The attire of the figure with a mace in Pl. X: 3 is not the usual tunic but a garment opening in front, somewhat resembling a worshipper's mantle but short, indicating the possibility that a foreign type of garment was represented. The number of examples of the common scene showing the figure with a mace facing only the suppliant goddess increases in Level III, Pl. X: 4, 5, 6 and 7. But the majority of this type of representation comes from Level II, Pl. XI: 1, 2, 3 and 4, Pl. XII: 2 and 3. Occasionally, small motifs appear between the two figures: Pl. XI: 1, 2, 3, 4 and Pl. XII: 2 and 3.

In the other representations in Level II, the figure and goddess are flanked by several columns of inscription, an element that became an important part of the seal design during the reign of King Ipalpiel II of Eshnunna. Subsidiary motifs are absent, with the exception of Pl. XII: 4. The sealings concerned are Pl. XI: 5, 6, 7, and 8, Pl. XII: 3 and 4, Pl. X: 9, 10, 11, and Pl. XII: 5 and 6. They were probably intended to be similar to the above-mentioned group, but, owing to carelessness in their rolling, most of the composition is missing and attention is therefore drawn to the remaining inscription.

The descriptions of the figure with a mace in the Harmal impressions and discussion of other published examples are relatively easy tasks compared with the attempt to determine the identity of the figure. Although he is the most common figure in Old Babylonian seals, aside from the suppliant goddess, scholars have not agreed who he is. Frankfort called him "the god with the mace . . . who may stand for a variety of deities. . . He may also represent the king, or even the king's statue."[109] E. Porada in *Corpus*, p. 52, states: "The fact that he usually wears a cap, like that of Hammurabi on his stele, rather than a horned mitre suggests that he is a king and not a deity . . . the general uniformity in the rendering of the figure may imply that it represents a king as such rather than any specific ruler." In *Sippar,* p. 105, she points out: "Perhaps the fact that the god with the mace usually stands at the left,

[107] For other examples of the figure with mace at the end of a devotion scene see *Louvre* II, Pl. 80, No. 19 (A. 371); *VR*, 380, 392; *Ur Ex.* 10, 505; *Cyl. Seals,* Pl. XXVI i, where it stands on a platform; *Danish Ntl. Museum,* 39; *Corpus,* 413; Moortgat-Correns, *Baghdader Mitteilungen* 4, Pl. 43, No. 65; J. Piet, *JANES* Columbia Univ. 2-4, pp. 30-36; *Bib. Nat.,* 219; and an unpublished seal in the Iraq Museum, IM 13554. Various gods are depicted as the main protagonists of these scenes, therefore the figure's presence is not limited to an association with the enthroned "deified" king.
[108] Other examples are *VR*, 343; *Brett,* Pl. VII 59.
[109] *Cyl. Seals,* 168.

which is not the honored side reserved for major deities who almost always appear on the right, indicates that the figure is human as a king in war-like attire."

Earlier, B. Landsberger suggested that he represents the god Ninshubur, while B. Parker rejects this theory[110] because Ninshubur should hold the staff of *sukkalu*-ship and not a mace.[111] R. Moorey, with regard to the terracottas and seals from Kish, associates him with the local god Zababa.[112]

It is a fact that the figure with a mace never receives an offering in Old Babylonian seals[113] with the exception of a unique scene where he receives an offering from a priest, and where he appears at one end in what may be a sacred marriage with a goddess.[114] It is doubtless the fact of the sacred marriage with a deity to which the royal figure owes the sacrifices usually reserved for gods. In one or another of the impressions from Harmal the figure with a mace can fall within any of the categories suggested above. As E. Porada suggested, the figure is definitely not a major deity. At the same time, it is difficult to call him just a mortal or a king. Although the later seals from Harmal, which are accompanied by inscriptions, normally state that the owner of the seals are the servants of King Ibalpiel II, it is possible that the king is represented here facing the suppliant goddess. But, in the earlier sealings from Levels IV and III, it would be more difficult to see him representing a king; a suppliant god, like the suppliant goddess, is more likely, particularly when he comes either before or after the worshipper.

In the earlier seal designs, the appearance of the figure with the mace before major gods may show him there either on his own behalf or as a representative of his people. He faces the goddess Ishtar in Pl. VI: 10.[115] Occasionally he faces the Sun God.[116] In fact, examples of the occurrence of this figure with various gods are too numerous to be cited here.

1.1.12. The Suppliant Goddess

The suppliant goddess is rarely discussed on her own in published seal catalogues. But her frequent occurrence in the Harmal impressions makes it necessary to devote a few lines to her varied functions. She appears on the earliest seal impressions, sometime leading a worshipper towards a deified king or an enthroned god, or standing behind the worshipper. Later her position changes: in Level III she appears predominantly facing the figure with a mace.

She always appears wearing a horned crown indicative of her divinity, though her posture in relation to major deities always points to her minor status in the hierarchy of gods. Her

[110] Parker, *Iraq* 37, p. 25, note 35 and Landsberger, *Ankara* 1948, no. 89 and note 226.
[111] *Ibid.*, No. 8.
[112] Moorey, *Iraq* 37, p. 83.
[113] *Cyl. Seals*, Pl. XXIX b.
[114] Unless the female worshipper facing the figure with a mace would be regarded as such in *Corpus*, 394.
[115] *Corpus*, 371-374; *Ashmolean*, 509; *Legrain*, 390 and 395. In *Newell*, 218, the suppliant goddess is facing Shamash and the figure with the mace is facing Ishtar. Here, at least, he and the goddess appear to perform a similar function.
[116] *Corpus*, 421, 422, 424; *Ur Ex.* 10, 479.

hair is worn as a chignon with long curls falling back across her shoulders. Her long garment consists of slightly overlapping tiers of pleats or flounces, and may have been made of linen and not of wool.[117] In some impressions, particularly from Level II, she seems to have a long tassel at the back of her dress ending with a counterweight for the necklace worn by this goddess during the Old Babylonian period.[118] She has been identified with the goddess Lama, an intermediary between gods and worshippers.[119]

1.1.13. The Nude Female

Few examples of the nude female occur in the Harmal impressions, usually as a filler motive. There are two from Level IV which, in fact, come from the same envelope (No. 34). One has been discussed already in relation to the Weather God, Pl. VIII: 1. In the other impression, Pl. X: 2, the nude female is the main figure, approached by the figure with the mace who is followed by a worshipper. In sealings of Level III, she appears twice behind the Sun God, in Pl. XIII: 1 and 2. In Pl. IX: 8 she is behind the god with a mace. A faint impression of her is seen on a tablet dated to Naram-Sin, Pl. XIII: 3. Her rather frequent appearance in the sealings of Level III probably indicates her popularity at this period. She is depicted only once in Level II, Pl. XIII: 9.

She is always naked and facing front, with her hands clasped under her breasts. Her hair is rather short, hanging down only to the level of her shoulders. This is in contrast to the rendering of Ishtar's hair which hangs down in locks to her elbows. The nude female is better seen in terracotta figurines of the same period.[120] Sometimes she looks as if she were wearing a choker.[121] Two almost life-size heads of clay statuettes, most probably representing the nude female, have been acquired recently by the Iraq Museum.[122] The coiffure of these heads is so similar to those on the terracotta figurines and differs so much from representations of female worshippers occasionally seen on Old Babylonian seals,[123] that it is possible to consider them as having belonged to statues of the nude female set up in temples. This would indicate that she was an important personage in Babylonian rituals. A seal in the Ashmolean Museum[124] depicts her standing on a pedestal, approached by a figure offering a goat. The

[117] Wiseman, *Iraq* 22, p. 166.

[118] See Spycket, *RA* 42, pp. 89-97. Basing her study on seals, sculpture and other evidence, she concludes that the necklace appeared first during the time of King Shu-Sin of Ur and continued until the Kassite period. It was worn by both the interceding suppliant goddess (her Figs. 3 and 4) and mortal women (her Figs. 1 and 2). Dales, *RA* 57, pp. 21-40, points to earlier examples from the Agade period. He also suggests that the terracotta figurines representing the tambourine players (his Fig. 10), wore such a necklace. It had a wide distribution in the ancient Near East. Yet, with all the new evidence, there is still uncertainty as to the precise name and significance of the necklace, though it must have been of great importance to those who wore it. Dales, relying on textual evidence, suggests that the necklace could have been of gold with lapis lazuli for the counterweight at the bottom of the cord. See also Maxwell-Hyslop, 1971.

[119] For her identity and role, see: *Cyl. Seals,* p. 160; *Corpus,* p. 55; Wiseman, *Iraq* 22, pp. 166-171; and Spycket, *RA* 54, pp. 73-84.

[120] Barrelet, 1968, Pl. XXXVIII.

[121] *Ibid.,* Figs. 389-395.

[122] Abbada, *Sumer* 30, p. 334, Figs. 1 and 2 (Arabic section).

[123] *Corpus,* 394, 568. *Diyala,* 913.

[124] *Ashmolean,* 479.

figure is wearing the flounced garment of the gods and is followed by a second god. This could support the suggestion that she played an important role within the temple.

1.1.14. Embracing Couple

Two seals from Level III, Pl. VII: 3, and Pl. XIII: 4, dated to Hammidashur, and both from the same archive,[125] depicted the figure of a goddess extending one hand to touch a worshipper. The worshipper extends both his arms, one to reach behind the goddess to touch her far shoulder, the other towards her waist. The subject seems always to be combined with other motifs, equally important to the overall composition of the seal. Pl. VII: 3 has the god with the scimitar facing the lion-headed monster.[126]

The peaceful atmosphere of the embracing couples, always represented by a mortal and a goddess, possibly illustrates the divine marriage.

1.1.15. Small Figure of a Worshipper

Certain impressions portray a small figure of a worshipper nude, or wearing a short and rounded cap-like headgear, between the figure with a mace and the suppliant goddess: Pl. X: 7, Pl. XII: 1, and Pl. XIII: 5. The figure is probably the image of the seal owner.

An impression from Level IV, Pl. XIII: 6, represents a small worshipper kneeling on one knee raising the right hand in front of his face while placing the left forward. A similar figure is among the filler motifs of Pl. XIII: 7. Both come from Level IV.[127]

A figure of a long-robed worshipper appears as a filler motif in Pl. XIII: 8.[128] Two small figures of long-robed worshippers are placed on either side of a large figure of a worshipper in P. XIII: 9.[129]

[125]See *Al-Hashimi.*

[126]*Louvre* II, Pl. 78, No. 26 (A. 314) has a similar composition: the god with the scimitar is facing a worshipper. *Diyala,* 906, has the demon behind the goddess. There is a great similarity in the execution and engraving of both seals, which are thus possibly the product of one craftsman. Frankfort, *Cyl. Seals,* Pl. XXIX b, was influenced by an inscription to suggest that the motif has a connection with the myth of Nergal "who forced an entrance into the nether world and took Ereshkigal for a spouse, " p. 170. He also conceded the impossibility of proving such a theory.

[127]Sollberger, *Iraq* 31, pp. 90-93, states that statues in the same posture were dedicated by worshippers to their gods and placed in temples, Pl. X. Porada, *Studies . . . Oppenheim,* 1964, p. 161, indicates that the figure was probably of Syrian origin as it appears frequently in Syrian seals of the early second millennium. Later examples in Old Babylonian seals are: *Bib. Nat.,* 154 and 232; *Cyl. Seals,* Pl. XXVII; *Newell,* 266; and *Yale,* 919.

[128] Other examples are *Corpus,* 427, 480.

[129] In a cylinder in the Walters Art Gallery, *Gordon,* No. 21, a small figure of a worshipper appears as a filling motif.

1.1.16. Old Assyrian and Other Peripheral Impressions

Among a number of envelopes at Harmal, a few have impressions typical of the Cappadocian seals belonging to the Old Assyrian group.[130] Pl. XIV: 1 and 2 have been dated to Sin-abushu and come from the same room.[131] Both are typical presentation scenes. The impression of No. 2 is on the envelope of a legal adoption document. Pl. XIV: 3, also of Sin-abushu date, shows a god holding an axe over his shoulder, a common motif in Old Assyrian seals[132] and a suppliant goddess is seen behind a worshipper. The style of this impression is that of the Late Old Assyrian group as emphasized by the face and headgear of the worshipper.[133] The rendering of the goddess' garment, with flounces giving the effect of being wrapped around the body, is uncommon. Another sealing of Late Old Assyrian style and of Sin-abushu's date is Pl. XIV: 4 from the archive of Nurshamash. This shows a suppliant goddess behind a worshipper who faces the Sun God and a nude hero in conflict with a lion. A star disc in the crescent is in the sky.

Another impression of Late Old Assyrian style is on an envelope from Harmal, Pl. XIV: 5 of Naram-Sin date: another adoption document. The impression is nearly complete and shows part of a suppliant goddess behind a priest with pail and sprinkler facing an enthroned god on a platform. In the sky there is a serrated vase and a star disc in the crescent. In a secondary group, a winged lion-demon and a lion attack, between them, a human victim.[134] Closely related in style is sealing Pl. XIV: 6, also from Level III, where a similarly rendered lion repeats the attack on a human victim. A suppliant goddess accompanies a worshipper of whom only a part is visible while the principal deity of the scene is missing. A serrated vase above a ball-staff appears between the goddess and worshipper.

Two unusual impressions are assigned to the reign of Hammidashur. Pl. XIV: 7 depicts a nude hero, wearing headgear resembling a furry cap, fighting a lion. The hero is attacked from the back by a lion-headed demon, whose wings are rendered with deep horizontal incisions. A second similar nude figure, holding a weapon, is being attacked by a lion-headed demon who brandishes a weapon over his head. The terminal consists of a snake. It is difficult to find a parallel for the style of this impression, with the exception of a haematite seal from Ishchali.[135]

The second seal, Pl. XIV: 8, is more perplexing. The impression is that of a row of men hand in hand, all nude but wearing a round cap-like headgear and performing a "war dance." The first dancer holds his left hand forward, the last dancer holds his right hand over his head. A small male figure appears in the empty space. Rows of men do indeed occur on seals from the Diyala,[136] but are of a different style or theme. The subject in Pl. XIV: 8 is a motif which

[130] Names as such by Porada, *Corpus*, Pl. CXXVII; Özgüç, 1968, Pl. IVA, Pl. X E,F.
[131] See *Suleiman*, No. 84.
[132] *Corpus*, 877; Özgüç, 1965, Pl. IX, 27.
[133] For this terminology, see E. Porada, "Karum" *RLA* 4, p. 383-389. For parallels for the worshipper's cap with vertically striated brim see Porada, *Corpus*, 865; Özgüç, 1968, Pl. XVII A.
[134] The style and particularly the carving of the lion is identical to a seal from the Kanish *Karum* Özgüç, 1968, Pl. XVII A. Related representations are *Corpus*, 880 and *Louvre* II, Pl. 95, No. 12 (A. 885).
[135] *Diyala*, 931.
[136] *Diyala*, 725, 741, 905, 935.

became popular later on Mitannian seals at Nuzi,[137] and frequently as a secondary motif on Syrian seals.[138] Its presence at such an early date may illustrate that a number of motifs, supposedly originating outside the Mesopotamian plain and occurring at a later period (e.g., the winged goddess of Pl. IX: 7) actually occur earlier than anticipated in Harmal. Two possibilities arise: either such motifs had developed in Mesopotamia, or there was intimate contact with the peripheral regions in which these motifs were already in use.

A typically Old Assyrian impression on the envelope of Pl. XIV: 9 was found in Level II, and has an Ibalpiel II date formula.[139]

The impression of a crudely carved cylinder of indeterminate style, Pl. XIV: 10, is also from Level II. A walking, apparently nude male figure is followed by a thin, tall nude figure holding, with one hand, what looks like the head of a human victim, while brandishing a weapon in the other. The lower part of a worshipper is visible; he is offering a kid to the Sun God. The impression is most probably an importation.

1.2. Tell Dhiba'i

The mound located 4 kilometers north-east of Tell Harmal is rectangular in shape and 300 by 150 meters in dimension. In 1947, a short season of excavation was undertaken by the Department of Antiquities and five levels of occupation were uncovered.[140] The two upper levels were shallow and occupied a very small area of the mound, so that when excavation was resumed in 1962 those levels were quickly cleared, particularly the temple area. Thus, Level V of the first season is Level III of the second season. A building complex on the northern mound was extensively dug. In addition, six hundred tablets were unearthed in the temple during the second season. Tablets were also discovered in a room adjacent to the temple and these may have belonged to the temple archive.[141] The date formulae found on the tablets (from Level V where most of the seal impressions were discovered) can be correlated with the date formulae from Harmal Level III, particularly with those belonging to the reign of Hammidashur.[142] Unlike the tablets from Harmal, which come from private houses and thus belong to individuals, the Dhiba'i tablets come only from the temple and the complex of buildings on the northern mound which may be an administrative center. They are mostly loan contracts. The ancient name of Dhiba'i has been identified as the city of Uzarzalulu, and its main deity is the god Lasimu. Both names are found in the inscription of the seal impression, Pl. XIX: 1, which occurs on several of the loan contracts attributed to the archive of the god Kash. Thus it has been suggested that Kash is another name for Lasimu.[143]

[137] Porada, 1947, Pl. XIX.

[138] *Corpus*, 944; *Louvre* II, Pl. 95, No. 28 (A. 901). A Syrian seal impression on an envelope from Sippar, dated to the time of Hammurabi, shows a row of soldiers, No. 209 Pl. XXXVII. The impression is also published by Buchanan, *Iraq* 33, Pl. II. It is later in date than the Harmal example.

[139] A very similar seal design is seen in *Bib. Nat.*, 262.

[140] Mustafa, *Sumer* 5/2, pp. 173-198.

[141] See *Karim.*

[142] Baqir, *Sumer* 5/2, p. 140. Also see *Karim.*

[143] Karim, *Sumer* 23, pp. 189-192.

Unfortunately the state of most of the envelopes is very fragmentary and they cannot be related to any of the documents, so that there is no definite dating. But as all the tablets come from Level V and have a date contemporary with Hammidashur, it is possible to assign the impressions, tentatively, to the same period.

The subjects on the impressions are limited. There are two main subjects most frequently represented: presentation and worshipping scenes; and, less frequently, contest and fighting scenes. Few seal impressions with individual subjects occur. These will be treated separately.

The presentation scenes can be divided into two groups. Pl. XV: 1, 9, 12, 13 and 14 were probably locally made while Pl. XVIII: 2 and 6, and Pl. XIX: 2 and 6 were imported from another center.

1.2.1. Worshipping and Presentation Scenes of the Local Group

There is a uniformity in the style and size of the impressions of this group. Most of them are small, normally showing a figure, perhaps an enthroned king, who wears what seems to be a round cap and a long garment and who holds a cup. His seat is a throne with a back and side decorated with a panel design, thought to represent a temple facade or doorway. Sometimes the "king" receives a worshipper with clasped hands held at the waist and wearing what looks like a round cap and long garment, occasionally with a fringe. This worshipper is usually followed by a second one, who raises his right hand in supplication in front of his face, while the left is held to his waist. Pl. XV: 1 and 2 are probably impressions of the same seal. They show a third worshipper wearing the same round "cap" as the person before him, but a short kilt in contrast to the long garments of the other two worshippers. The first worshipper in this impression is bareheaded. In Pl. XV: 3 a single horn is depicted emerging from the forehead of the seated figure and may have been added as an afterthought. Pl. XV: 4 is most perplexing. The seated figure and inscription are identical to b 1, but in b 2 the second figure has acquired long hair and what looks like a horn to its cap. It is possible that these were later additions to the seal. In Pl. XV: 5, only one worshipper is depicted. The space that was occupied elsewhere by a second worshipper is filled here with two rampant lions holding between them a staff, the head of which is missing A similar motif occurs as a filler motif in another impression from Dhiba'i, Pl. XV: 6.[144]

Behind the worshipper in Pl. XV: 7 is a hero, wearing a short kilt and what looks like a round cap, restraining a lion. Behind the "king" on the same impression is a nude figure holding a scimitar and resting its top on the ground. Owing to the fragmentary state of the impression, it is difficult to determine if the figure is that of a god or a mortal.[145]

[144] Two similar lions without the standard are seen in a fragment from Tell Harmal Pl. I: 13. It comes from Rm. 520, Level III and is tentatively dated to Hammidashur, the same period as the sealings from Dhiba'i.
[145] *Corpus,* 305 has a nude figure of a warrior wearing what seems like a round furry cap, carrying a similar scimitar. The warrior faces an enthroned "king". The style and the secondary motifs point to a Diyala origin for that seal. A similar warrior is depicted on *Nippur I,* Pl. 115, No. 13.

A number of impressions which belong to the local group depict the interceding goddess leading a worshipper towards an enthroned "king" or "god": Pl. XV: 8, 9, 10, 11, 12, 13, 14, 15, and 16. The goddess in No. 8 wears a flounced garment revealing one bare shoulder, derived from goddesses of the Ur III period,[146] although such a garment continued to appear occasionally on seals of the Isin-Larsa or early Old Babylonian period, c 2000-1800 B.C. Part of the flounced garment of the enthroned deity holding a cup is visible. Such a cup is rarely seen in the hands of gods.[147] The unusual symbol of a double crook with a single staff appears between the goddess and the worshipper.[148]

The terminals of both Pl. XVI: 15 and 16 are a double lion-headed club; zigzagged tassels fall down on both sides of the spiked pole; the spikes may represent metal coils wound round the pole. The tassels do not occur elsewhere but are seen on a crescent standard from Tell Asmar.[149] The engraver of Pl. XVI: 16, must have misunderstood the mace and engraved a five pointed meaningless object instead.

An uncommon feature is the presence of a full-sized figure of the bow-legged dwarf in Pl. XVI: 1,[150] and beside him is a giant lion-headed scimitar. Such enlargement of the symbols occurs frequently during the early Old Babylonian period as well as towards the end of the period. The earlier occurrences were frequently in terminals[151] while those of the late period appeared in the main scenes.

The small impression of Pl. XVI: 2, shows a worshipper wearing what looks like a round cap leading a second worshipper towards the enthroned deity. An interceding goddess was probably intended, but the seal is poorly carved, which accounts perhaps for the neglect in depicting the horned crown. The same feature occurs in a crudely cut seal from Tell Asmar.[152]

In Pl. XV: 10 the shallowness of the impression fails to demonstrate details of the design. Thus, the interceding goddess probably wears the flounced garment, while the "king" is enthroned on a fleeced stool as in Pl. XV: 9. The worshipper has a descending line of fringe marked by short oblique incisions along the outer edge of his garment. Parallels to this impression can be found among the seals from the Diyala sites.[153]

The garments in sealing Pl. XVI: 3 differ generally from the sealings previously discovered. The suppliant goddess wears a long pleated skirt and the worshipper a plain robe without fringe. The sealing shares with the foregoing ones both style and secondary motifs.[154]

[146] *Corpus*, 277, 292.
[147] *Ibid.*, 309; also in *Louvre* II, Pl. 76, No. 21 (A. 244), the deity wears the flounced garment of the gods and the round cap-like headdress of mortals.
[148] For a similar motif, see *Louvre* II, Pl. 81, No. 8 (A. 381), identified by Delaporte as a double crook. This is possibly a development from the two crooks appearing together. *Ibid.*, No. 12 (A. 384).
[149] *Diyala*, 771.
[150] Usually he is depicted on Old Babylonian seals in miniature as a subsidiary motif.
[151] *Corpus*, 293, 305; *VR*, 296, 311.
[152] *Diyala*, 750.
[153] *Ibid.*, Pl. 68; also *Yale*, 700.
[154] For related cylinder seal designs see *Diyala*, 728, 752.

Three impressions from fragments of envelopes are unusual in their workmanship but fall within the subject range of the seals at Dhiba'i. Two illustrate devotional scenes with a suppliant goddess behind a worshipper facing the enthroned "king," Pl. XVI: 4 and 5. The third, Pl. XVI: 6, depicts a row of three goddesses facing a similar personage. The impressions are unique, as no parallels have been found for the style and manner of carving, and they are possibly the work of one craftsman. The similarity does not stop at the style alone, but also includes the main figures and many of the secondary motifs. The enthroned "king" wears the round cap and a fringed garment. The suppliant goddesses in the three impressions wear single-horned crowns, and flounced garments that leave one shoulder bare. This is unusual, as flounced garments of suppliant goddesses cover the shoulders and hang down below the elbow of the left hand. The face is carved with a prominent, pointed chin, the neck is long, and the hair is gathered into a large bun. The engraver may have misunderstood the purpose of the ribbon, or he could have portrayed a local fashion.[155]

The procession of the three goddesses in Pl. XVI: 6 is most uncommon in Old Babylonian sealings. The composition suggests derivation from Agade seals, but no other element on these seals supports such an imitation. The influence is probably attributable to Old Assyrian seals where rows of deities file in procession towards an enthroned "king."[156]

The ball-staff and vase appear in the three impressions, the vase having a ribbed neck and circular body. A lizard appears in Pl. XVI: 4 and 6 with a disc in crescent in the sky. The terminal in Pl. XVI: 4 and 5 consists of two columns of a faint inscription, the cuneiform sign for the god Adad appearing in both of them. No parallel was found for the motif of a snake eating a lizard in Pl. XVI: 6. However, both symbols occur frequently, but separately, on the Diyala seals.

Although the composition of Pl. XVI: 7 is completely different from Pl. XVI: 4, 5 and 6, it is probably the product of the same workshop. In addition to the similarity in the rendering of the figures, the "king" and the two worshippers are wearing similar garments. The crescent, mongoose, star, vase, ball-staff and snake are all seen in one or another of these three sealings. What makes the impression on Pl. XVI: 7 distinctive are the large-scale figures of two nude wrestlers behind the "king." They are wrestling with interlocked arms.[157] Such pairs appear as subsidiary motifs in Old Babylonian seals, usually as small figures. They are frequently in a posture of dancers holding hands, lifting one leg which crosses the knee of the second leg and which each figure holds up with the second hand. Sometimes the figures are seen with their bodies crossing each other,[158] probably in a wrestling posture, which has a long tradition dating back to the Early Dynastic period.

[155] A possible parallel to the ribbon falling from the hair could be seen in the impressions from Tell al-Rimah: Daley, Walker, Hawkins, 1976, Pl. 109, Nos. 13 and 16.

[156] For examples see *Corpus,* 851 and 860; *Gordon,* Pl. V, No. 34.

[157] The same motif in different execution is seen in *Louvre* I, Pl. 53, No. 27 (D. 131). To judge by the rendering of the figures' faces and headgear, that seal is probably a product of the Diyala region.

[158] For the dancers see *Corpus,* 555 and 556; *Louvre* II, Pl. 79, No. 25 (A. 341) and *Yale,* 730, an impression from Harmal. For the bodies crossing, see *Cyl. Seals,* Pl. XXIX M; *Bib. Nat.,* 132; and also the fragment from Tell Harmal No. 68.

Devotional scenes where no deity is represented, but where the object of worship is a divine symbol, occur in several envelope fragments at Dhiba'i. Pl. XVII: 1 and 2, show the shoot or frond of a "date palm" in a vessel as the center of the composition.[159] The subsidiary motif of Pl. XVII: 2 is a kneeling figure;[160] while in Pl. XVII: 1, part of a bow-legged manikin is visible. The manikin appears again in Pl. XVII: 3, where the central symbol is a bird balancing a crescent standard over its back.[161]

Single figures performing various functions, perhaps participating in ritual scenes, also occur. In Pl. XVII: 4 a nude attendant stands behind the enthroned "king" wearing a belt with a tasseled knot at the back. He carries a "whisk" in his right hand and has a folded "towel" hanging over his left lower arm.[162]

An exceptional subject is seen in Pl. XVII: 5. There are two worshippers, one behind the other, with a spear standing between them. A third figure, which follows, is a nude male wearing the same cap-like headgear as the worshippers, but holding a bow. This is a rare motif whose exact meaning is difficult to determine.[163] However, the symbolism of this impression seems to be of a warlike nature.

Several individual features may be noted in various sealings from Dhiba'i. In Pl. XVII: 6 from Level V, a male figure with his hair shown like a round furry cap, has a short kilt with triple belt knotted at the back. Before the figure stands a worshipper with similarly arranged hair. In Pl. XVII: 7 a nude figure with a pointed chin walks behind a priest with a pail and sprinkler. Both priest and nude figure wear knotted tasselled belts. Pl. XVII: 8 is similar to a seal from Harmal Pl. XVII: 9 depicting a row of men. These impressions form a distinct group sharing many similarities: the rendering of the figures and their postures; several figures are usually depicted on a seal; and there are often columns of inscription. Moreover, they are the product of one region, most probably the Diyala. A seal that is similar to the Dhiba'i group comes from Ishchali.[164]

The seal impressions of the local group discussed so far, (Pls. XV, XVI and XVII) share characteristics that could suggest that they were locally manufactured. One such characteristic feature is the crude engraving of most of the figures. The "king" and worshippers all wear the identical headgear resembling round caps, except Pl. XVI: 7. The "king" is always enthroned on a

[159] Seen on a seal from Tell Asmar, *Diyala,* 716; also *Corpus,* 314, where there are worshippers on either side of a stylized tree. The symbol is derived from Ur III, the stele of Urnammu being the best example.

[160] For a similar figure from the Diyala, *Diyala,* 999.

[161] In *Louvre* II, Pl. 79, No. 23 (A. 339), the bird balancing a standard is a terminal on a seal where the double lion-headed club is the center of worship. *Corpus,* 313, though damaged, may depict such a bird. *Newell,* 148, is an Old Assyrian seal depicting two dancers balancing a crescent standard with a bird underneath (head obliterated). The significance of the symbol is perplexing, as it is not known that a bird is associated with the Moon God.

[162] For the interpretation of the scene, see J. Börger-Klähn, *ZA* 64, pp. 235-240. In general, attendants behind deities are common in Old Babylonian seals. An attendant is seen behind the "king" on Tell Harmal No. 46a, dated to Hammidashur. The priest with pail and sprinkler is shown behind Shamash on an impression from Harmal, No. 11a, Pl. V, dated to Sin-abushu and thus earlier than the Dhiba'i impression.

[163] *Corpus,* 382, has a male figure carrying a bow over his shoulder and assisting the conquering king in subduing an enemy. Also *Yale,* 901.

[164] *Diyala,* 900. Another such seal is *Corpus,* 553, possibly originating in the Diyala region.

panelled-type stool, exceptions being Pl. XVI: 4, 6 and 7, and Pl. XVII: 4. The smaller and secondary motifs are limited. There is always a crescent in the sky and a roughly cut eight-pointed star. The ball-staff and vase appear on a number of the impressions.[165] Other motifs which occur less frequently are: the mongoose in Pl. XV: 2 and Pl. XVI: 5 and 7. A bird appears in Pl. XV: 14 and Pl. XV: 3 No. 3 has a plant as a terminal (a rare motif in Old Babylonian sealings). In Pl. XV: 14, there is a scorpion between the goddess and the worshipper. But the most distinctive feature in the style of the local Dhiba'i repertory is the occurrence of two worshippers in one scene, each with a different gesture. Also distinctive is a noticeable absence of the suppliant goddess.

1.2.2. Elaborate Presentation and Worshipping Scenes Not of Local Origin

Presentation and worship scenes in impressions discovered in the same level (V) at Dhiba'i are obviously not of the same style, but correspond more closely to impressions from Tell Harmal and may have come from the same source. Unlike the locally manufactured group, where uniformity of style and subject of the seals are apparent, the majority of the impressions are independent in character, particularly with regard to their subject matter (Pls. XVIII and Pl. XIX).

The suppliant goddess appears in most of the seals wearing the multiple-horned crown and the flounced garment, as on many of the Old Babylonian seals from other sites. The "king" wears the round cap with upturned brim and is enthroned on a fleeced-type stool. He is sometimes seen receiving a bareheaded worshipper, Pl. XVIII: 1 and 2, a theme derivative from the Ur III repertory which occurs in several impressions at Harmal: Pl. II: 5 and 6 from Level II; Pl. II: 2, 9 and 10 from Level III, with a Hammidashur date.

The terminal in Pl. XVIII: 2, a hero mastering a bull, can be compared to the impression from Harmal Pl. II: 8 (Level III, Hammidashur date), where the hero is overpowering a lion in a similar posture. There are other similarities in the secondary motifs, such as the drill holes scattered in the field, the sun disc in crescent, a mongoose in Pl. XVIII: 2 and a monkey in Harmal Pl. II: 8. They point to the popularity of certain motifs on contemporary seals that probably originated from one source. Both Pl. XVIII: 3 and 4 show female worshippers with the characteristic gesture with which they are represented in Old Babylonian seals.[166] The right hand is extended forward just above the waist (a begging gesture?) instead of being raised in front of the face, as in the gesture of worship made by men. Unusual is the counterweight necklace worn by the female worshipper in Pl. XVIII: 4 and not, as expected, only by the suppliant goddess. This figure may, therefore, be that of a priestess. A small figure of an eagle with outspread wings, which appears between the female worshipper and the "king" in Pl. XVIII: 3, is an uncommon motif in Old Babylonian seals but occurs frequently in the Ur III examples.[167] Impression Pl. XIX: 5 has a grooved border similar to

[165] Their meaning and actual identification are not certain. See Porada, *Syria* 43, pp. 254-256 with reference to the interpretation of these objects by Özgüç, 1965, p. 57.

[166] *Corpus,* 394.

[167] For such examples, see *Cyl. Seals,* Pl. XXV f., *Corpus,* 284; an eagle appears also on an early Old Babylonian seal of "Isin-Larsa" style, *Ibid.,* 309.

Harmal Pl. IV: 3 from Level III,[168] and made with the ridges of the metal cap of the original cylinder or with the seal's imitation of the cap in stone. The bent object between the "king" and worshipper in Pl. XVIII: 5 may represent a snake.[169] The crook in this impression is very rare in Diyala seals; at Dhiba'i it is the only example.

The nude attendant with frontal face and three pairs of curls holding a flowing vase is only seen in sealing Pl. XIX: 6 at Dhiba'i. Furthermore, there is only one enthroned god holding the rod and ring represented at Dhiba'i, Pl. XIX: 7. The panelled throne is carefully and elaborately rendered here in contrast to the impressions of local seals. In fact, this impression can be compared with the best seals of the Old Babylonian period. The same can be said of a seal impression on a jar sealing of Pl. XVIII: 8 representing the goddess Ishtar, which depicts every detail of her garment and weapons. The suppliant goddess facing her is treated with the same accuracy, and her three-row necklace with a counterweight is correctly shown. Both impressions clearly belong to imported seals.

In addition to the official temple seal, Pl. XIX: 1, showing a god in an ascending posture,[170] a second impression, Pl. XIX: 2 which is also from the temple archive, shows a god in a similar posture, but this can be Shamash. There is an attendant in this impression holding a rather unusual curved object. It is difficult to say whether or not this is a purification whisk as in Pl. XVII: 4 but as he appears to be standing on a dais this is perhaps right.

Pl. XIX: 3 is a reconstruction from several fragments depicting the "hieros gamos" in two separate scenes. In the first the female participant is drinking from a vessel although the tube cannot be seen.[171] The second scene depicts the couple on a bed. This subject was frequently depicted in terracottas, but rarely on seals of the Old Babylonian period. Two other seals from the Diyala region, dating to the Early Dynastic period, however, show a similar subject. The seal from Tell Asmar illustrates the couple on a bed with a vase placed at the head of the bed.[172] In addition to the main subject, the Early Dynastic cylinder shares with the Dhiba'i impression the vessel and the scorpion, indicating that the symbolism of the ritual survived unchanged from the earlier period.[173] The second seal from Khafajeh[174] is crudely engraved, but it also has both scenes depicted.[175]

[168] For actual examples of seals with caps imitated in the stone of the seal see *Louvre* II, Pl. 74, No. 18 (A. 191); *Newell*, 185 and 186.

[169] The cylinders *Legrain*, 323 supposedly showing a "crooked stick" and *Danish Ntl. Museum*, 38, have two heavy, engraved parallel curved lines similar to the one on the Dhiba'i sealing No. 131c, Pl. XIX.

[170] Identified as the god Lasimu, i.e. "runner" by M. Matoušova, *Sumer* 22, pp. 115-117.

[171] The motif is a popular one on Old Babylonian terracotta reliefs. One actually comes from Dhiba'i. M. Mustafa, *Sumer* 5/2, Pl. VI, No. 8; also Barrelet, 1968, 527 from Telloh; Opificius, 1961, 612. Several plaques were discovered at Kish, probably in association with the goddess Ishtar: see Moorey in *Iraq* 37, Pl. XXV; examples are also seen in *Nippur I*, Pl. 137. The scene is represented in a few seals that can be attributed to the Old Babylonian period: *Louvre* I, Pl. 52, Nos. 21 (D. 148) and 20b (D. 149). Also the terminal in a seal with a presentation scene, which depicts the female bending and drinking with a straw from a vessel: "Collection de cachets et Cylindres Orientaux," auction catalogue, Paris, Hotel Drouot. Vente: 14. V. 1962, Pl. 4, No. 46.

[172] *Diyala,* 559; *Yale,* 458 for similar symbolism.

[173] Also see a Syrian seal from Latakia, though later in date than the Dhiba'i impression, in H. Seyrig, *Syria* 32, pp. 38-41, Pl. IV, No. 3. It shows a scorpion and a goat in the field. A priest is blessing the couple, as in the Asmar seal.

[174] *Diyala,* 340.

[175] A later Syrian seal also has both acts depicted: E. Porada, in *Vorderasiatische Archaologie . . . A. Moortgat*

Two seal impressions among the Dhiba'i collection are clearly Old Assyrian. Their presence is not unusual, as Old Assyrian impressions occurred also at Harmal. Pl. XIX: 4 is similar to Harmal Pl. XIV: 1 and 2 while Pl. XIX: 5 belongs by its style to the same group as Harmal Pl. XIX: 109. In the Dhiba'i impression a scorpion man is depicted sitting on a platform. It was difficult to trace a similar example, thus its significance is uncertain.

1.2.3. Contest and Fighting Scenes

The two pairs of contestants in Pl. XX: 1, 2, 3 and 4, and the group of a frieze of heroes and animals in Pl. XX: 5, 6 and 7 are represented at Dhiba'i. Unlike the locally-made presentation scenes which can be distinguished from the general iconography of the Old Babylonian seals, the contest scenes at Dhiba'i fall within the typical Old Babylonian repertoire.

The hero in profile holding a bull in an inverted position in Pl. XX: 5 and 8 is similar to the figure on Pl. XVIII: 2 which suggests a common source of manufacture for all of the original cylinders of these impressions. The terminal of Pl. XX: 8 is unusual, consisting of a cluster of drill holes, a recumbent goat with head turned backward, a bird, a monkey, and a seated dog.

The winged lion-griffin is depicted in several impressions. In Pl. XX: 5 it attacks a goat; in Pl. XX: 10 and 7 a kneeling human figure is the victim, who is placed on a platform in Pl. XX: 11. Impression, Pl. XX: 12 illustrates an uncommon subject: the male figure is no longer a passive victim but a hero confronting the demon who looks as if it has submitted to his powers.

Some of the characters in the animal frieze from Dhiba'i can be paralleled by impressions from Harmal, both in style and subject. One example is the hero restraining his victim, a lion, by bending one knee and pushing it toward the back of the animal: Dhiba'i, Pl. XX: 5 and 7; Harmal Pl. I: 4. The winged lion-headed griffin is also illustrated in the Harmal impression. Another similarity is an animal victim standing between two assailants. Dhiba'i, Pl. XX: 5, shows a lion between two heroes, while in Pl. XX: 7 the lion's assailants are a hero and a bull-man. In Harmal, Pl. I: 8, the bull is the victim of a lion and a bull-man.

The unity in subject and style of the fighting scenes from Harmal and Dhiba'i makes it probable that they were bought and manufactured at one common source.

1.3. Conclusion

There were at least two or more sources for seals prevalent in the Diyala region during the earlier part of the Old Babylonian period. One indicates a local industry and the other possibly shows a wider acquaintance with styles found throughout the region.

1964, Pl. 33:4. For the significance of these motifs see: *Cyl. Seals*, p. 75; H. Seyrig, *Syria* 32, pp. 38-41. Also the discussion by J. R. Cooper, "Heilige Hochzeit" in *RLA* 4, pp. 259-269 and *Yale*, p. 200.

1.3.1. Local Style

The local seals are plain in style, mostly portraying presentation and worshipping scenes, best illustrated by the group of impressions from Dhiba'i. These do not stand alone, but are linked to seals from other Diyala sites: Asmar, Khafajeh, and Ishchali where a collection of seals discovered in the Isin-Larsa-Old Babylonian levels may have been the product of one workshop. These seals mostly illustrate a goddess wearing a single-horned crown, and a distinctive, long pleated garment with two crossed straps to hold it over the shoulder. The goddess leads a worshipper with his hair indicated by a round cap, sometimes of the furry type. The worshipper also wears a long garment, normally with a fringe of irregular scratches that looks as if it were detached from the garment. This treatment of the fringe may be due to poor craftsmanship. Both the goddess and worshipper approach an enthroned "king," who is always seated on a panelled stool. The best example comes from Tell Asmar[176] which probably provided the model for other examples.[177] The quality of the work is poor, as the engraver sometimes neglected to show the horns of the goddess, and her chignon is reduced to a simple straight line.[178] The faces of all participants are completely stylized; a vertical groove for the face and neck and two or three lines for the nose and mouth.[179]

Like the impressions from Dhiba'i, the secondary motifs are limited. A rampant lion[180] or a goat[181] appear as terminals, sometimes crudely cut and stylized in common with the rest of the seal's engraving.[182] Another motif for a terminal is a crescent standard,[183] frequently with tassels,[184] and seen already in an impression at Harmal, Pl. II: 3. One crescent standard has two long zigzagged ribbons,[185] similar to the ribbons on the double lion-headed clubs from Dhiba'i, Pl. XV: 15 and 16. As at Dhiba'i, most of the sealings have a crescent in the sky and an irregularly cut star.[186] A bird also appears,[187] as in Pl. XV: 14. The lizard is also common as a filler motif.[188] The ball-staff is depicted in several sealings, but none shows a vase, which is unusual for an Old Babylonian seal.[189] By far the most frequent motif that occurs in these seals as a filler or a terminal is the snake.[190] The frequent presence of this representation conforms with the religious beliefs of the area.[191]

Most of the local seals come from Tell Asmar, indicating that the workshop was probably located there. The local industry at Dhiba'i, discussed above, has many characteristics in

[176] *Diyala,* 723.
[177] *Ibid.,* 522, 539, 734, 775.
[178] *Ibid.,* 737, 740, 775.
[179] *Ibid.,* 539, 711, 775, 953.
[180] *Ibid.,* 732, 742, 953.
[181] *Ibid.,* 743.
[182] *Ibid.,* 739, 750, 751, 766.
[183] *Ibid.,* 740.
[184] *Ibid.,* 713, 717.
[185] *Ibid.,* 771.
[186] *Ibid.,* 522, 717, 737, 744, 775.
[187] *Ibid.,* 764, 775. On 732 and 737 the bird is crudely cut.
[188] *Ibid.,* 539, 764, 948.
[189] *Ibid.,* 539, 711, 781, 782.
[190] *Ibid.,* 522, 739, 711, 713, 717, 723, 732, 734, 737, 953.
[191] Delougaz, *OIP* LXIII, 1952, p. 121, Pl. 129.

common with the Asmar group. These include the basic subject of a goddess leading a worshipper to an enthroned "king," and the limited iconography of the secondary motifs. But examination of the details reveals that each belongs to a different workshop. The Dhiba'i craftsman, albeit working with stylized strokes, took more care in showing some of the facial details. The ball-staff and vase occurred more frequently, while snakes are fewer.

It is difficult to give an exact date to the Asmar group. But as the Dhiba'i impressions are tentatively dated to Hammidashur, the Asmar seals should be contemporary.

Apart from the fact that poorly worked seals are found at most of the Mesopotamian sites, the seals and impressions from the Diyala region share little with those that were discovered in southern Mesopotamia. For example, none of the seals discovered at Ur [192] depict a goddess wearing a pleated garment with straps across her chest. In fact, at Ur a distinctive coarse type of local work is apparent.

Similarities in the rendering of the seals and some of the secondary motifs of Tell Dhiba'i can be found with those of Susa.[193] At Susa only one figure faces the "king," but both wear fringed garments and the hair is indicated by what looks like a round furry cap. But the fringe of the garment that projects away from it, seen in most of the Diyala seals and in Dhiba'i, Pl. XV: 9 and 12, is absent from the Susa seals. Fringes indicated by scratches pointing away from the garment are a common feature in the seals of the Old Assyrian group.[194] Impressions of this group occur at both Harmal, Pl. XIV: 1, 2, and Dhiba'i, Pl. XIX: 4.

1.3.2. Regional Style

The local industry of Asmar and Dhiba'i is less apparent at Harmal, where only a few impressions occur, Pl. II: 3 and Pl. XXI: 1, 2, and 3. This suggests perhaps that these impressions were made with cylinders bought from the production centres of Dhiba'i or Asmar.

There is little change in the subjects and style of both Levels IV and III at Harmal. In historical terms this is the reign of Hammidashur of Level III. Typical inferior works formerly assigned to Isin-Larsa are depicted on the same tablets with impressions considered to belong to the mature style which developed and flourished in Babylonia proper (ex: Harmal tablet No. 3, Pl. 1).[195] In comparison with Diyala seals of Asmar and Dhiba'i, Harmal had a different source for the bulk of its seals, particularly those from Levels IV and III. The impressions from the Nurshamash archive should be included here due to the similarity of the iconography and date.

It is difficult to determine if a local industry was set up at Harmal itself. Several differences in the details of the seals may suggest more than one origin. For example seals like Pl. I: 1,

[192] *Ur Ex.* 10, Pl. 23 and Pls. 31-33.
[193] *Susa,* Pls. 166 and 167, the seals are a local product; Nos. 1831-33 are possibly the product of the Diyala region.
[194] *Corpus,* 846 and 849.
[195] *Porada,* pp. 155-162.

Pl. III: 8, 6, 3 and Pl. XXI: 4 and 5 are of inferior workmanship when compared to seals such as Pl. II: 8, Pl. V: 1, Pl. VIII: 7 and Pl. XIII: 4.

A change in style and subject made its appearance at the end of Level III during the reign of King Naram-Sin of Eshnunna, and developed later in Level II. Although inscriptions occur in a number of impressions from as early as Level IV, their proportion is small. For instance, only four impressions from the archive of Nurshamash have inscriptions, and one has the panel of Dingir Shamash Dingir A-A, Pl. XXI: 6, 7, 8, 9, 10 and 11, an inscription popular particularly on seals of the early Old Babylonian period. In Level IV at Harmal four impressions have this inscription: Pl. II: 1, 5, 12 and possibly 7; Pl. III: 7, 8, and 9. There are fewer inscribed impressions in Level III, with the exception of the envelope No. 54 (Pl. II), dated to Naram-Sin, where four out of the six impressions have inscriptions. This illustrates the beginning of a trend in emphasizing the importance of inscriptions. When subjects became more uniform and difficult to distinguish, showing mostly the figure with a mace and the suppliant goddess, particularly in Level II, inscriptions were vital. In addition to the practice of rolling primarily the inscribed part of the seal with only a small part of the scene with figures visible, there was a change in the practice of using the seals. Tablets had become smaller and had no envelope. Therefore the impressions are seen on the tablet itself, where for the most part only one seal was rolled. Another innovation was the appearance of seals with caps made of metal (gold), and decorated with filigree work as seen in Pl. XI: 5 and 9, the seal of King Ibalpiel, and Pl. XI: 10 and 11. Their early presence at Harmal should be noted, as the practice is thought to have evolved in the post-Hammurabi era and to have become characteristic of the Kassite seals.

Few seals at Harmal were definitely imported from outside the boundaries of Babylonia. Pl. IV: 4, discussed above, seems to have come from the upper Euphrates. There are a number of impressions of the Old Assyrian seals, Pl. XIV: 1, 2, 3, 5, 6, and 9, indicating the extent of trade and foreign contact that the Diyala region had with the north and east. Impressions illustrating particular subjects that appear later on Mitannian seals make their appearance at Harmal, e.g., the winged goddess in Pl. IX: 7 and the dancing men of Pl. XIV. These examples are too few to indicate which way the influence was moving, i.e. whether they are Mesopotamian motifs in origin, or whether the Hurrian infiltration already made itself felt. The long beaked bird on Dhiba'i No. 131.i (Pl. XVII: 3) is another of these motifs that became popular later on at Nuzi.

CHAPTER 2

AKKAD

2.1. Tell Ed-Der and Sippar

Sippar and its suburb Tell Ed-Der are the sites chosen to represent the Old Babylonian seals from Babylonia proper—Akkad as opposed to Sumer. This is due to the particular abundance of tablets discovered at these sites during the last century which have enriched many a museum in Europe and America.[1] Sippar flourished during the early part of the Old Babylonian period and seems to have been an agglomeration of several small cities,[2] each inhabited by one of the western Semitic tribes that were settling in Babylonia at that time. During its early history, local rulers were in control of the city and the surrounding countryside. But in the reign of Sumulael, King of Babylon (c. 1880-1845 B.C.), Sippar came under the domination of Babylon and remained as such until the end of the First Dynasty of Babylon.[3]

The majority of the texts come from the archive of the temple of Shamash, the main deity of the city. A large number belong to the *nadītu*, a certain class of priestesses associated with the temple. The majority of the texts are dated to Sin-muballiṭ (c. 1812-1793), Hammurabi (c. 1792-1750) and Samsuiluna (c. 1749-1712).

Tell Ed-Der is located 116 miles southwest of Baghdad and only a few miles from Sippar. A sounding was undertaken there in the spring of 1941 by the Antiquities Department of Iraq.[4] The mound was fortified with a *libin* wall. The sounding also revealed a number of private houses and a public building. Several levels were excavated. Those that concern us are Levels II and III where more than 300 tablets were unearthed, primarily from the public building. One archive has been identified as belonging to an individual by the name of Anum-pisha.[5] Most of the tablets were slightly earlier in date than the documents of the *nadītu* priestesses from Sippar. Although they are fewer in number, they are of special importance as the private archive is important as a contrast to the Sippar collection, which belongs to officials and temple personnel.[6]

[1] Numerous publications of the texts from Sippar have appeared since the end of the last century and still continue. The latest book is *Harris,* which gives a comprehensive history of the city. The Department of Archaeology, University of Baghdad, is currently excavating at Sippar: W. al-Jadir and Z. Rijab, "Preliminary Report on the Excavations of Baghdad University at Sippar," *Sumer* 39, pp. 79-122 (Arabic).

[2] Ibid., pp. 10-14.

[3] Ibid., pp. 1-9.

[4] Baqir and Mustafa, *Sumer* I, pp. 37-54.

[5] Tell Ed-Der has been identified as Sippar-Amnanum, one of the important suburbs of the city of Sippar, *Edzard*, p. 19. Two groups of tablets were published. The first belongs to the earlier part of the Old Babylonian period, mostly dated to Apil-Sin and Sin-muballiṭ, the second group to Samsuiluna and Ammiṣaduqa.

[6] A Belgian expedition has been excavating at Tell Ed-Der for the last few years. About 2000 tablets and fragments have been discovered during the season of 1975. Many have sealings, and therefore, their publication would be of great interest.

The impressions from Tell Ed-Der and Sippar will be discussed in one section as the chronological and stylistic aspects of the glyptic of both belong together and complement each other.[7]

2.1.1. Impressions of the pre-Hammurabi Era

The earliest seal impressions come from envelope No. 182 (Pl. XXII) dated to Buntahtunila, a local ruler of Sippar who was a contemporary of Sumulael of Babylon, and represents the best in the glyptic of the early Old Babylonian period.[8] It is clear from these impressions that at Sippar the seal cutter was meticulous in showing all the details of the garments and facial characteristics In addition, the variety in subjects and imaginative representation of motifs indicate that during this period seals were made to individual order within the prevailing tradition. Such individual traits consisted of Ishtar holding a multiple weapon (Pl. XXII: 1), made of alternating maces and double lion-headed clubs, and of the spread-winged eagles on either side of her head, a derivation from Ur III glyptic.

The throne of the god in Pl. XXII: 2 is carved in the form of a bull instead of the usual panelled stool. The seal is expertly executed with minute and intricate details. Entirely original is the drawing of the nude hero in profile showing his three side locks. The tail of the lion which he grasps passes between his legs, while his hands wind round the neck of the lion to catch its forward legs. The result is a finely integrated composition. In Pl. XXII:3 the Sun God is depicted enthroned on a high-backed throne associated with him elsewhere (discussed above with the Harmal seals). Pl. XXII: 4 belongs to a group of seals which were popular during the earlier part of the Old Babylonian period. The conquering god or warrior is here assisted by a second warrior. The warrior god also appears on Pl. XXII: 5; Pl. XXIII: 1 and 2.[9] Complex mythological scenes which include lion-headed demons, the god with club, and the conquering god were popular on seals. Possibly, these were the result of a continuous tradition from the Agade period in the depiction of complex mythological scenes.[10]

The terminal of Pl. XXII: 4 consists of two registers depicting fighting scenes; a similar terminal appears in Pl. XXIII: 3.[11] Pl. XXII: 4 belongs to the same group and depicts an heirloom impressed on a Hammurabi tablet. Compositions of fighting scenes are imitations of Early Dynastic

[7] The majority comes from the Louvre: *Louvre* II. The texts are published in *ThD*. The unpublished impressions included in my catalogue belong to the texts published by Figulla as part of the British Museum cuneiform texts publications Vol. 47, here designated by *CT*, and a few impressions belong to Vol. 45. I am grateful to the authorities in the Western Asiatic Department for their kind permission to let me study the collection.

[8] Already published by Waterman, *AJSL* 29, pp. 202, No. 2184A.

[9] For a parallel to the garment of the god, see *Ashmolean*, 467.

[10] An impression on a tablet dated to the later years of Hammurabi (Buchanan, *Yale Library Gazette* 45, No. 9; *Yale*, 796) belongs to this group of impressions and was most probably manufactured between the time of Sumulael and Apil-Sin. All the motives can be paralleled by earlier impressions from: Sippar, Nos. 182.2 and 5; Der, 154a; from the archive of Nurshamash No. 152; and *Porada*, p. 159, Fig. 1.

[11] A similar terminal could be seen on No. 152a of the archive of Nurshamash. An impression depicting a similar terminal on a tablet dated to Sabium, thus falling within the same group, Waterman, *AJSL* 29, p. 202, No. 2189A. Similar seals already published which can now be dated to the early part of the Old Babylonian period are: *Legrain*, Nos. 330, 441.

examples[12] with innovations by the Old Babylonian seal cutter consisting of double register scenes. Terminals in two registers with antithetical subjects also occur: Pl. XXIII: 5 and 6. Similar motifs are also seen at Larsa, Pl. XXXIX: 2 and at Ur, Pl. XLII: 9 within the framework of the early Old Babylonian period.

At Tell Ed-Der, the god Adad is represented three times: Pl. XXIII:7, 8 and 9. He is shown in association with the winged-lion dragon, nos. 7 and 8. Pl. XXIII:7 is most unusual, as the deity seems to be sitting on a throne over the back of the bull. The line between the bull and the figure with mace is possibly a scratch in the original seal, because there is neither precedent nor parallel for the figure with mace leading a bull.

In Pl. XXIV: 1 a goddess, possibly Gula, holding the ring and staff is depicted standing on a platform balanced over two crouching dogs.[13] In the same impression, the platform of the attendant is similar to the panelled thrones of the gods. The dog balancing a stick occurs on two more impressions, Pl. XXIV: 2 and 3. As the goddess Gula was depicted on an impression at Der, the dog here may symbolize the goddess.

Offering and devotional scenes, described in the catalogue as presentations, occur most frequently. Normally the main feature is an enthroned "king," Pl. XXII: 1, 4, 5, 6; Pl. XXIII: 1, 2, 5; Pl. XXIV: 7 and 8. The worshipper is normally bareheaded and clasps his hands in front of his chest (except Pl. XXIV: 2).

The high quality of the engravings of these seals conforms with the rest of the glyptic from this region in contrast to the examples from Dhiba'i and Asmar. Such presentation scenes disappear by the time of King Sin-muballiṭ, with the advent of the workshops at Sippar.

2.1.2. Impressions of Sin-muballiṭ and Hammurabi Date

During the time of Sin-muballiṭ a certain style in the manufacture of seals seems to have arisen at Sippar, or Akkad in general.[14] Two workshops, if not more, can be distinguished.[15] Both portray similar subjects, but differences in the cutting and treatment of the figures reject a master-apprentice situation, and point to different artistic centers during the prime of the Old Babylonian period. These workshops are numbered I and II here.

2.2 Workshop I

Workshop I is distinguished by the triple undulating lines used to render the flounces of the garments. During this period Old Babylonian seal composition abandoned the unified scene in favor of a juxtaposition of a number of divine figures apparently unrelated to one another. This development is particularly noticeable at Sippar. The engraving is bold and every detail is carefully and neatly rendered, indicating that skillful artists were at work.

[12] *Cyl. Seals*, Pls. XII-XIV.
[13] Deities standing on two antithetical animals, are seen on other contemporary impressions. *Porada*, p. 159, Fig. 1. The motif becomes popular later. This will be discussed with the Sippar workshops.
[14] Unfortunately, as the finds are limited to Sippar comparative material for the rest of central Babylonian is lacking.
[15] Porada already called attention to one workshop in *Sippar*, p. 111.

The subjects of the seals are very limited; the majority are either offering or devotional scenes. The main deities represented are Shamash, Ishtar, Ea and the god with mace.

2.2.1. The God Shamash

Shamash is represented in his normal ascending posture holding a saw. Furthermore, the figure of a god who holds a staff, a ring, or a ring and staff can also be identified as Shamash.[16] He is represented enthroned on a panelled stool in an imprint from the time of Sin-muballiṭ, Pl. XXV: 5. This is unusual because normally when Shamash is enthroned his throne is high-backed, Pl. XXII: 3, or engraved with mountain scales, Pl. XXV: 1. A similar representation is seen on a relief of this period, where Shamash, with ring and staff, is enthroned on a panelled stool. In the seals he is always portrayed wearing the flounced garment with undulating lines, regardless of his posture, which differs from his appearance in the impressions from Workshop II. The association of the secondary motif of a kneeling bull with Shamash, suggested by E. Porada with regard to the seals from Sippar, is probably correct.[17] The bull is depicted in front of Shamash in Pl. XXV: 2, 3 and 8.

2.2.2. The Goddess Ishtar

Ishtar is usually depicted holding her weapons with one foot placed on her lion, Pl. XXV: 6.[18] In one case she is enthroned holding the double lion-headed club, with her face in full view.[19] The goddess also appears in a new posture which became a popular motif on the Sippar seals: standing on two reclining lions and holding the double lion club in one hand, while the other hand is placed by her waist. In these scenes she wears the flounced garment with undulating lines and not her usual pleated skirt, Pl. XXV: 7, 8 and 9.[20] The center mace of the club is omitted on No. 7.

2.2.3. The Water God

Ea is represented frequently. This again is a feature peculiar to Sippar.[21] He seems to have had a sanctuary in the city but he played a minor role in the Sippar pantheon.[22] He is depicted mostly on impressions from the Hammurabi period, usually seated on a throne placed over a goat-fish with his feet resting on a second goat-fish. He holds a vase when enthroned, in Pl. XXVI: 1 and 2,[23] and a flowing vase when standing, Pl. XXVI: 3. In No. 2 a nude attendant with three pairs of curls, and holding a vase, faces him. He holds the vase not in the accustomed position close

[16] Ibid., p. 104.

[17] A. Moortgat, 1969, Pl. 209; *Sippar*, seal XXVII.

[18] *Sippar*, seal XXVII.

[19] Ibid., seal XXXIII.

[20] The depiction of Ishtar is similar to the god with club-mace who only wore the pleated skirt when in an ascending posture (see section on the sealings of Tell Harmal).

[21] Few sealings from other sites portray the water god, and appear to represent mostly seals made to order.

[22] *Harris*, p. 148.

[23] Ranke, *BE* VII, No. 7.

to his chest but away from him and toward the personage with an offering before him. This, again, is a feature distinctive of Sippar.[24]

Damkina, Ea's wife, is also represented on the seals of this class. She does not appear on seals from the previous period, and seldom on seals from other sites. This shows that her presence was another special feature of the Sippar workshops. She appears fullface, with hands clasped to her chest, holding a flower vase and wearing a single-horned crown.

In Pl. XXVI: 7, the flowing water is not visible but fish are seen swimming upstream.[25]

2.2.4. The God with Mace

The god with the mace and the figure with the mace are both present in the impressions, though not as frequently as in Workshop II. The god with mace is distinguished by the chignon hairstyle reserved for the gods, and a tall cap with a narrow brim. The brim is possbily a stylization from a single pair of horns, Pl. XXVII: 1 and 2. The frequent occurrence of this headgear at Sippar is probably due to contact with the Upper Euphrates where, according to the mural of Zimrilim, a tall domed cap was in fashion.[26] The god in Pl. XXVII: 3 wears a similar headdress, as well as the minor deity in Pl. XXVII: 5. The god with mace shares with the figure with mace the long swinging beard.[27]

2.2.5. Minor Deities

Minor deities, in addition to the suppliant goddess, are depicted on most of the seals and are represented in most of the impressions. This is another characteristic of the Sippar seals. Two new goddesses are introduced. One is shown fullface, wearing a flounced garment that covers both her shoulders. She has one hand extended forward and usually stands behind the deity. Her gesture is that of introduction, but her identity is difficult to determine. She can be the consort of Shamash.[28] She appears in seals, Pl. XXVI: 6 and 7; Pl. XXVIII: 1.[29] She is depicted facing the figure with a mace in Pl. XXVIII: 2, behind the water god in Pl. XXVI: 6, standing opposite another goddess in Pl. XXVIII: 4, and behind a worshipper in Pl. XXVIII: 4.[30] Another fullface minor goddess wears a crown with a single pair of horns, a long garment having one shoulder bare, and both hands clasped at chest level. Her function is not clear apart from accompanying other

[24] Ibid. Also *Sippar*, seal XXXIII, and an unpublished seal in the Louvre, AO 25.518, show two nude attendants holding their vases in a similar position.

[25] The water goddess is depicted on a seal in the Louvre, AO 25.518, where the flowing water from her vase meets those flowing from the vases of the two attendants standing by her sides to touch the heads of the fish-goat. On Ranke, *BE* VII, Pl. IV, No. 7, her long skirt is engraved with wavy lines as if symbolizing water. She also appears on Ranke, *BE* VII, Pl. IV, and with a similar garment on the statue from Mari, *MAM* II.

[26] *MAM* II, Pl. XI.

[27] It is tempting here to suggest the seal No. 1769 in *Susa* could be the work of a Sippar craftsman because of the close parallel of many of the motifs. Certainly *Susa*, No. 1757, is an import from Sippar.

[28] *Harris*, p. 150 mentions that Aja shared the temple with Shamash and was a favourite of the *nadītus*.

[29] Also, *Sippar*, seals XVIII and XXXIV.

[30] The impression belongs to a high priest of the temple, which could explain the worshipper's flounced garment.

gods and worshippers, Pl. XXVI: 7; Pl. XXVII: 1 and 7. The goddesses, as well as minor gods, exhibit several types of crowns, excluding the multiple horned crown worn by main deities, the suppliant goddess and even various goddesses depicted on the Sippar impressions. One crown has a single pair of horns with a round top, Pl. XXVII: 7. The goddess' hair is short, falling down the neck to end with locks.[31] Another type of horned crown has a square top, also seen in the sculpture of the period.[32]

Minor gods, or gods with no definite identification, also occur. One is seen with head in profile and hands clasped at chest level, behind an enthroned god in Pl. XXVII: 2 or behind a worshipper in Pl. XXVIII: 5, where he is wearing the long plain garment of mortals. The god offering a kid in Pl. XXVII: 5, Pl. XXVIII: 6 and 7, a motif favoured by the Sippar workshops, is unusual.[33] A god whose identity cannot be determined in Pl. XXVIII: 8 holds what appears to be a plant, and wears a particular type of long skirt, with the flounces rendered in zig-zag lines.[34] Also seen frequently on the Sippar seals is the bull-man offering a kid to the Sun God, Pl. XXVII: 6; Pl. XXVIII: 1, 9 and 10.[35] The nude attendant with three pairs of curls is depicted serving in offering scenes, but only in association with Ea, Pl. XXVI: 1, or his wife Damkina, Pl. XXVI: 8.[36]

2.3. The Styles and Characterizations of the Sippar Workshops

Apart from the manner of rendering the flounced dress of the deities on the seals of Workshop I at Sippar, no other iconographic feature is unique. The subjects of Shamash, Ishtar, Ea, the filler motifs, etc., are all part of the Old Babylonian repertoire. At Sippar it is the combination, arrangement, and the repetition of the same subjects that make these seals distinctive. Most of the seals have two separate scenes, and both can be equally important, such as on Pl. XXV: 8, Pl. XXVI: 7, Pl. XXVIII: 3, 5, 7, and Pl. XXXIV: 1.[37] Although the enthroned deities are derived from earlier glyptic tradition, at Sippar they were probably an imitation of existing works of art, most notably the Code of Hammurabi. Several gods appear in the same impression, sometimes without a worshipper as in Pl. XXV: 9, and Pl. XXIX: 1. They seem to be a procession of gods even on seals where a worshipper is depicted, Pl. XXVIII: 3 and 4, and Pl. XXXIV: 1.[38]

The repetition of the same figures on the seals can also be seen with secondary motifs, which the Sippar cutters sometimes fancied for filling the vacant spaces between the larger figures. The most popular of the motifs were a reclining bull, usually placed by the head of the main figure, composite fish creatures (goat-fish, lion-fish, and fish-man), and severed heads (man, bull-man, and the hero with three pairs of curls). Other motifs which also occur include the nude female

[31] A similar crown and headdress is seen on the statue of the water goddess from Mari, *MAM* II, Pl. V.

[32] The bronze statuette of the four-faced goddess from Ishchali, OIP LX, Pls. 79 and 80.

[33] Ranke, *BE* VIII, Pl. IV, No. 7. The figure mentioned could represent a priest wearing the garment of the gods. Because of that and the horned crown, it is included here.

[34] The impression here is faint, but a similar god is seen in *Corpus*, 391; *VR*, 503. He also appears on an impression dated to king Zimrilim, where he holds a large branch behind the water god. *Syria* 37 (1960), p. 215, fig. 1.

[35] *Sippar*, XXVII.

[36] Ranke, *BE* VIII, Pl. IV, No. 7.

[37] *Sippar*, seal XXVII; *BE* VIII, Pl. IV, No. 7.

[38] Ibid., Ranke.

always a filler motif), the small male figure, a fly, the bow-legged dwarf (manikin), a goat sitting on a knoll, a cow suckling its calf, and crossed animals.

The continuous recurrence of the same subjects and secondary motifs through the reigns of Sin-muballiṭ, Hammurabi, and Samsuiluna suggests that a workshop of father and son, or perhaps of three generations, as suggested by E. Porada,[39] is most probable. A conservative tradition was observed as little variation is seen. This was possibly maintained by having certain patterns offered to the customers, and may account for the lack of interest in assigning a space for any inscription. It is also possible that the first seal cutter was illiterate[40] and did not include any inscriptions in his products. The owner may have added the inscription later, possibly by another cutter. With the exception of Pl. XXVI: 1 and Pl. XXIX: 2, where proper columns of inscriptions are depicted, all the others have the inscriptions scattered in the field. In Pl. XXIX: 3 part of the inscription is within two columns and the rest in the field, suggesting that the columns were an addition. This applies also to the case of the box of inscription of Pl. XXV: 9, Pl. XXVIII: 3, and Pl. XXXIV: 3, where it can be noticed that the inscription was undoubtedly imposed over the seated lion in Pl. XXXIV: 1.

A certain development in the style and occurrence of some motifs seemed to have appeared at the end of Hammurabi's reign. These became more apparent during the time of Samsuiluna; for example, the frequent occurrence of the cow suckling its calf which appears first on an envelope from the 35th year of Hammurabi, Pl. XXVI: 2, and Pl. XXVIII: 5.[41] The two crossed bulls appear on a tablet of the 43rd year of Hammurabi on No. Pl. XXIX: 4. Both motifs appear in one impression from Samsuiluna's 10th year, Pl. XXVIII:3.

The change in style with an increase in the use of the drill becomes more frequent with Samsuiluna.[42] Drill holes are left unmasked on certain parts of the composition, such as the chignon of the god, Pl. XXIX: 5, and Pl. XXXIV: 1. The beard of the god in No. 1, and the fringes of the worshipper's garment in Pl. XXIX: 6 are also executed with drill holes.[43]

Finally, the repetition in the work of the craftsmen of Workshop I did not mean that they were unable to engrave motifs other than these already discussed. Individual seals were probably also made to order. This proves the excellent abilities of those craftsmen, as seen on Pl. XXIX: 7, which depicts the unusual figure of the bull-eared god stepping on a lion-headed demon, and Pl. XXVII: 3 which shows the conquering god with the multiple mace (made from seven double lion-headed clubs). All the subjects—the bull-eared god, the conquering god, and the unusual multiple club—occur on seals of the early Old Babylonian period. Clearly the cutter was copying an old example.[44]

It is difficult to show the extent of this workshop, due to the limited number of accessible tablets. However, judging from the frequency of the impressions, its rise possibly occurred at the end of

[39] *Sippar*, p. 111.
[40] This is only a supposition as there is no evidence to show whether the seal cutters in Babylonia were literate or not.
[41] *Sippar*, seal XXXIV.
[42] Discussed in detail later when the drill style becomes the prevailing one at Sippar.
[43] The standard with drill holes on *JNES* 16 (1957), pl. 195, Fig. 3, dated to Hammurabi's 43rd year. Drill holes for the fringe of the worshipper's garment to be seen on *Bib. Nat.*, 236; *VR*, 503.
[44] The goddess Ishtar holds a multiple mace club in *Sippar*, No. 182.2, Pl. XXII.

Sin-muballiṭ's reign. It then flourished during the Hammurabi period and decreased in frequency by the time of Samsuiluna with the workshop's fall in popularity, possibly caused by the rise of new drill and simplified styles.

2.4. Workshop II

We turn now to Workshop II, which had already been established and flourishing from the beginning of Sin-muballiṭ's rule. It was most probably extending the tradition of the established seal cutting industry already in existence at Sippar.

The main difference that distinguishes Workshop II from Workshop I is the treatment of the deities' garments: Workshop II rendered flounces in the traditional manner of straight incised lines. The subjects depicted are similar to those of Workshop I with some motifs occurring more frequently, such as the bull-man wrestling with the nude hero or lion, Pl. XXX: 1-8, and Pl. XXXI: 1. Impressions resembling those of Workshop I are few during the time of Sin-muballiṭ, Pl. XXV: 5, Pl. XXVI: 4, Pl. XXVII: 5, and Pl. XXX: 3. They are more numerous during the reign of Hammurabi when Workshop I was more active, Pl. XXV: 4, Pl. XXVIII: 8, Pl. XXX: 8 and 9, Pl. XXXI: 2-5. During the reign of Samsuiluna there is a continuation of the same motifs, but, as is the case with impressions from Workshop I, drill holes are more in evidence.

Among the secondary figures on impressions of Workshop II and absent from Workshop I is the nude priest with pail and sprinkler, Pl. XXV: 5, Pl. XXXI: 6, 7 and 8. It should also be noted that occasionally small figures of priests or other male figures are placed one above the other, Pl. XXXI: 7 and 9.[45]

Although columns of inscriptions are present on a large number of impressions from Sippar, some imprints of Workshop II show the inscriptions added after the original engraving of the seals, Pl. XXVII: 5, Pl. XXX: 9, Pl. XXXII: 1 and 2. As with Workshop I, however (see above), this practice is so constant that it may have been customary.

Seals likely to have been made to order also occur among the impressions of Workshop II. Pl. XXXII: 1 is a clumsy result of an attempt to depict a fullface, bearded god. Pl. XXV: 1 and Pl. XXXII: 3 show the embracing couples mentioned in connection with the Tell Harmal seals. In addition, Pl. XXV: 1 shows part of the Sun God enthroned on a high-backed seat decorated with mountain scales. The motif in Pl. XXXII: 4 of a fish speared by a trident is unusual. Its significance is difficult to determine but, judging by other examples, the trident was probably held by the god.[46]

At the beginning of Sin-muballiṭ's reign, seals with columns of inscription were numerous (in fact, a large number of the impressions on the tablets depict only the inscribed part of the seal). The common pattern shows either the god with mace or the figure with mace with the inscription consisting of three or four columns. During the earlier part of the period a worshipper was

[45] *Sippar*, seal X.

[46] A crescent standard ending at its base with three hooks, held by a god is seen on *Brett*, 68; *Danish National Museum*, 60. Collon, 1975, No. 135, p. 73, depicts an enthroned deity holding a similar standard with a fish hooked to it.

sometimes placed beside the inscription, Pl. XXXII: 5 and 6. The Sun God occurs in both impressions, and No. 6 has a bull-man holding a sun disc standard.[47] During the reign of Hammurabi a suppliant goddess stands on either side of the inscription, Pl. XXXII: 7 and 8.

Two suppliant goddesses on either side of the figure with mace occur on Pl. XXXIII: 1. A similar subject occurs on an impression from Tell Ed-Der dated to Sabium, Pl. XXXIII: 2. The flounces of the garments are rendered with undulating lines. Their manner of execution differ from the undulating lines of Workshop I; they resemble more the seals of the Ur III period. A similar rendering is seen on another impression dated to the reign of Apil-Sin.[48] The engraving of Pl. XXXIII: 1 resembles Pl. XXXIII: 3. Both have the filling of a small human figure: male in No. 3 and female in No. 1. Seals with identical figures have been found at Nippur[49] and Tell Harmal,[50] both belonging to women. Another seal from Mari[51] can be compared to Pl. XXXIII: 4. The large number of this type of seal on the tablets of Sippar suggests that the Nippur, Harmal, and Mari seals are the work of the Sippar Workshop II.

The figure of a female worshipper occurs on several impressions: Pl. XXV: 5, Pl. XXXIII: 5 and 6.[52] The documents come from the *nadītu* archives and thus it is not surprising to see them represented, nor that a large number of the inscribed impressions belong to *nadītus*. The female worshipper is distinguished by her headdress, with the hair gathered into a bun and secured with a band running round the forehead. She wears a plain garment with a short fringed shawl. One hand is folded underneath the garment while the other is extended forward. The female of Pl. XXXIII: 5 wears a necklace with a counterweight band already seen on a female worshipper at Tell Dhiba'i, Pl. XVIII: 4.

2.5. The Drill Style of the Late Old Babylonian Period

At the end of Hammurabi's reign, there was an apparent and gradual change in the style and subjects of the seals. The traditional motifs continued to be engraved, but a new fashion of cutting, which used drill holes for certain details, was adopted. Examples include the fringed border of the worshipper's garment in Pl. XXIX: 6 and the shoulder articulation of the Sun God in Pl. XXXI: 1.

The increase in the use of the drill during the reign of Samsuiluna is accompanied by a tendency to engrave taller and larger figures. In Pl. XXXIV: 1 the drill holes are used not only to indicate the fringe of the god with folded hands, but also the curls of his beard, a detail not otherwise designated. This figure is unusual as he is shown wearing the fringed robe of a mortal.[53] Another unusual aspect is the figure of the worshipper carrying a curved weapon over his shoulder. The god standing on two couchant bulls (or lions) with a crescent on his crown, belongs to an earlier

[47] Seen also on *Sippar*, seal XVI.
[48] *Porada*, p. 159, Fig. 1.
[49] *Nippur I*, Pl. 112, No. 14.
[50] IM 54659.
[51] *MAM* II, Pl. LVI, 32.
[52] Ranke, *BE* VI, I; Pl. IV, No. 7.
[53] A god in a similar attire is depicted in *VR*, 498. The seal is probably from Sippar and could also be of the same date.

period.[54] Tall figures are also noticeable in Pl. XXXI: 1 and Pl. XXXIV: 2. On the latter a small figure, probably the nude hero with flowing vase, is depicted, with the water rendered by drill holes. Plain drill holes were used in executing the locks of the nude hero-attendant in Pl. XXVI: 5 and Pl. XXXIV: 3. The sun disc is also executed by drill holes in the former.[55] A cluster of such holes in Pl. XXXIV: 4 is placed between the god with the mace and the suppliant goddess.[56] Its meaning is difficult to determine, particularly in combination with the three stars above. It is possible the design was carved according to the requirements of a particular customer.

In Pl. XXXIV: 5, the arms and legs of the figure with the mace are entirely rendered by drill holes. At first glance, the work may be regarded as an unfinished seal, but the evidence provided by the earlier examples of the drilled style suggests that this was intentional. Drill holes also form the pole of a stylized lion-headed club in Pl. XXXIV: 6. The impression of Pl. XXXV: 1, though partly damaged, shows the lower part of a suppliant goddess whose garment is stylized by horizontal lines in a fashion which also occur on an impression from Ur, Pl. XLIII: 8.[57]

The old traditional subjects continue to be depicted, but during the post-Hammurabi period the Weather God becomes increasingly popular, Pl. XXXIV: 7, Pl. XXXV: 2, 3 and 4. His normal posture of resting one foot on a reclining bull is seen in Pl. XXXIV: 8, or standing on a bull in Pl. XXXV: 3. He is facing another deity in Pl. XXXV: 5. This motif of two gods closely facing each other, particularly involving the Weather God, becomes a characteristic feature in post-Hammurabi glyptic. On these impressions a second distinctive development is apparent, i.e., the simplified representation of a crown with a single pair of horns resembling a side-brimmed hat as seen in Pl. XXXIV: 1. Similarly, the triple-horned crown may be represented by a series of drill holes bordering the crown, Pl. XXXV: 6.[58] Both stylizations become increasingly popular during the reigns of Ammiditana and Ammiṣaduqa.

Beside the conventional scene, Pl. XXXV: 7, probably a later addition, shows in the space reserved for the inscriptions the god Martu placing one foot on a gazelle. He is wearing one type of the "sun hat" crown, and the flounces of his garment consist of barrel-shaped grooves.[59]

[54] *Porada*, Fig. 1. The God on *Brett*, 67 has a similar crown. It is also seen on another Nippur impression of Ammiditana date No. 255.B.

[55] Porada, *JNES* 16, Fig. 3, dated to the end of Hammurabi's reign, shows a bull-man holding a disc standard executed with drill holes.

[56] The platform on an impression from Ammiditana's 11th year consists of four rows of drill holes indicating the popularity of the technique in forming any motif. Buchanan, *Yale University Gazette* 45, No. 16; *Yale* 999.

[57] A large number of seals already published in many museum collections depict this characteristic feature, the fringed garment indicated only by the horizontal ridges with a central wedge. It would be safe to date this class of seals to the reign of Samsuiluna.

[58] The method is clearly shown on *Ashmolean*, 547 dated to Ammiṣaduqa.

[59] The occurrence of such a figure on an impression dated to Samsuiluna helps in dating a number of published seals: *Danish National Museum*, 80; *Louvre* II, Pl. 83, No. 23 (A.456).

2.5.1. The Reign of Abi-eshuh

During the reign of Samsuiluna's successor Abi-eshuh, there is a noticeable decrease in the use of seals on written documents.[60] Although available examples are few, they indicate a continuation in the style and subjects of the seals with the drill extensively used as in Pl. XXXV:8 and 9.[61] A seal dated to the reign of Abi-eshuh shows a suppliant goddess facing an inscription with the outstanding absence of the horned crown usually worn by the goddess.[62] Instead, her hair is dressed in the fashion of the female worshipper seen on Old Babylonian seals, and she also wears a plain garment. The manner of execution is in line with the deterioration and simplification in the seal industry of the age. This type of suppliant goddess continues to be seen on seals in association with columns of inscriptions to the end of the Old Babylonian period, as attested by a number of dated seals and impressions.[63]

2.5.2. The Reigns from Ammiditana to Samsuditana

By the reign of Ammiditana there is a noticeable divergence from the traditional style of the seal industry at Sippar. The seals become smaller in size, resulting in cramped scenes with tiny figures. They are executed to a large extent by a drill. However, the older style of the Sippar workshops, in which drilling was restricted to details, continues to be found on tablets of this time. Whether such cylinders were currently manufactured or whether they were old seals in continued use cannot be determined. The impressions of tablet, Pl. XXXVI: 2, 3 and 4, are most probably the work of one craftsman which provide a good example here.

The new standard or staff with seven globes in Pl. XXXVII: 1 and 2 makes its appearancc.[64] The actual significance of this symbol during the end of the Old Babylonian period is not certain. It has been suggested that during this era it replaces the saw of Shamash.[65] The "sun hat" type of crown continues in use, Pl. XXXVI: 3. The crown becomes more simplified as a circle above a horizontal incision, Pl. XXXVII: 4. The Conquering God occurs in Pl. XXXVII: 1, 4 and 5. A single-horned crown in profile is probably intended for the Conquering God on No. 254.b, while a crescent crowns the "hat" of the second god. Over-indulgence in the use of the drill is seen in Pl. XXXVII: 2, where the features of the figures are hardly recognizable.

Subjects and technique continue unchanged through the reigns of Ammiṣaduqa (as seen also by the impressions of Tell Ed-Der) and Samsuditana.[66]

[60] Compared to the early part of the Old Babylonian period, few tablets have seal impressions in the post-Hammurabi epoch.

[61] Buchanan, *Yale Library Gazette* 45, Figs. 14 and 15; also *Yale*, 992-996.

[62] *VR*, 494.

[63] Dated in the time of Ammiditana are: *Louvre* II, Pl. 117, No. 4 (A.567); *VR*, 495, and *Ashmolean*, 551b. Dated in the time of Samsuditana is *Gordon*, Pl. IV, Fig. 26. This seems to be the only motif that prevailed until the Kassite period.

[64] *Sippar*, XXVIII.

[65] *Cyl. Seals*, p. 153. He does not give any reason.

[66] There are no impressions of Samsuditana's reign from the collections on which I have worked, but those published in *Yale*, 1021-1029, support this conclusion. A recent article by Evelyn Klengel-Brandt on seal impressions from Babylon dated mostly to Ammiditana and Samsuditana show clearly the popularity of the use of the drill in the execution of the motifs: *Altorientalische Forschungen*, 10, pp. 65-106.

The fine seals of the workshops of Sippar virtually disappear, with complete replacement by the drill technique. Motifs such as the Conquering and Weather Gods are the most popular. The Weather God is shown standing on a bull in Pl. XXXVII: 6, 7 and 8, or stepping on a reclining bull in Pl. XXXVII: 9.[67] The symbol in the hand of the Conquering God of Pl. XXXVII: 10 is possibly a distortion of the staff with seven globes.[68]

The gradual change from the styles of the Sippar workshops at the end of the Hammurabi period to the drilled style dominating the post-Hammurabi dynasty in Babylonia may have developed in Akkad if not at the city of Sippar. It is possible, though not likely, that the increased contact with the west may have played a role in the change of the styles. Syrian style impressions are found on the Sippar tablets, Pl. XXXVII: 11 and 12. A few Syrian motifs also appear on proper Sippar impressions, Pl. XXXVII: 10.[69] Sippar impressions from Workshops I and II occur at Mari,[70] and drilled style impressions appear at Alalakh.[71] Moreover, figures of the Old Babylonian repertory were included in Syrian examples, i.e., the use of the suppliant goddess,[72] or the enthroned god wearing the flounced garment with triple undulating lines.[73] Seals of this type illustrate that the Babylonian influence on the glyptic art of Syria was stronger than otherwise.

[67] Ibid., 1004.
[68] Ibid., 1020. Buchanan calls it a "bouquet tree." The impression is a good example of the drill style where holes were left exposed wherever possible.
[69] *Sippar*, XXIV.
[70] *MAM* II, Pl. LIII, 246: Pl. LVI, 287.
[71] Collon, 1975, 43, 45, 129, 133, 135.
[72] *Corpus*, 931.
[73] Ibid., 910.

CHAPTER 3

SUMER

The third region to be discussed and compared with the previous two (the Diyala and Babylonia proper), is southern Babylonia or what was known as Sumer. Impressions are from the two sites Ur and Larsa. Involved is the period when Larsa was in control of a large area of the south, particularly the city of Ur.[1]

3.1. Larsa

The impressions illustrated in this section belong to the group of business documents acquired by the Louvre at the end of the last century.[2]

There is a uniformity of subject and continuity in style from the earliest recorded impressions from Larsa. It is evident that here the industry was of a high standard, and the engraver's art equalled the best Babylonian work. This is apparent from the well known seal dedicated to the god Nergal for the life of Abisare (ca. 1905-1895 B.C.) and his own life by Puzur-Ninkarrak, a seal cutter or *bur-gul.*[3] If it is correct to associate the double lion-headed mace with Ninurta, then the god on the seal is not the god addressed in the inscription. In style it derives from the Akkadian tradition, though pointing to the fact that master craftsmen at Larsa were able to produce the finest works of that art. The impressions on tags dating to the reign of King Sin-iddinam are the earliest known from the site of Larsa.[4] The subjects conform to the glyptic of the period as scenes of worship with an enthroned "king" appeared (see Porada, figs. 2 and 3). The nude priest behind the enthroned "king" can be paralleled in examples from Tell Harmal, Pl. V: 5 or Dhiba'i, Pl. XVII: 4.

The impressions of Porada, figs. 6 and 7, are too fragmentary to allow a reconstruction of the rest of the scene. Most probably the figure with the mace would be standing at the end of a ritual scene or facing a major deity (e.g., Shamash or Ishtar), in conformity with his appearance on contemporary seals from Harmal, Nos. 4b and 42a (Pl. II). On the examples from Larsa, dated to King Warad-Sin (ca. 1834-1823 B.C.), he also appears as a secondary figure participating

[1] For a comprehensive history of the area see Hallo and Simpson, Chapter III.

[2] I am grateful to the authorities of the Louvre for permitting me to draw the impressions. The texts have been published in *Jean*. The date formulae were published by F. Thureau-Dangin, "La Chronologie de la dynastie de Larsa," *RA* 15, pp. 1-51. The impressions are described in *Louvre II*, with few illustrations. Not all the impressions mentioned in the catalogue have been drawn. This was due to the fact that many of the impressions were faint and some I have failed to see.

[3] Porada and Basmachi, *Sumer* 7, p. 67.

[4] Porada, pp. 155-162.

in a ritual scene in Pl. XXXIX: 3 and facing the goddess Ishtar on Pl. XXXVIII: 1 and Pl. XL: 6. He is standing on a platform decorated with a scale (mountain) pattern in Pl. XL: 5, a motif that was popular in southern Babylonia.[5]

Although the standard of engraving is high, the subjects depicted are very limited and restricted mainly to presentation and devotional scenes, with the exception of a fragment from the 9th year of Warad-Sin, illustrating a frieze of fighting figures, Pl. XXXIX: 10.

3.1.1. The Goddess Ishtar

The deity most frequently illustrated in the impressions from Larsa is the goddess Ishtar.[6] The characteristic skill of the Larsa craftsmen is shown in the meticulous rendering of the details of Ishtar's attire. Even the bracelets on her arms were depicted in Porada, fig. 12 (unfortunately the impression is only part of a small fragment). In Pl. XXXVIII: 2 the quivers, straps and pleated skirt are all carefully illustrated and so are the double lion-headed club and the lion underneath her foot (unlike the stylized examples from Tell Harmal). Facing the goddess is a rare example of a female worshipper (similar to Dhiba'i, Pl. XVIII: 3 and 4).

Three impressions from the reign of Rim-Sin, Pl. XXXVIII: 3, 4, and 5, illustrate a fullface goddess wearing a flounced garment and standing on two kneeling lions. Although she carries no weapons, she is most probably Ishtar, because of her association with the lion. These three seals look like the product of one engraver.[7] In addition to the goddess, they share a number of subsidiary motifs: a bull-man in contest is seen in both Nos. 3 and 4; Nos. 4 and 5 depict a kneeling bull by the head of the goddess; and a lion-fish is seen in Nos. 3 and 5.

Impressions in Pl. XXXVIII: 6, 7 and 8 depict Ishtar in her normal posture of placing one foot on her lion but, instead of holding the club-mace, here she holds a mace only. This seems to be a distinctive feature of Larsa (no other examples have been found at other sites). Pl. XXXVIII: 9 depicts a goddess holding a mace. The goddess Ishtar is probably represented here, though the rest of her symbols are absent.

3.1.2. The God Shamash

The god Shamash is also represented but not as frequently as Ishtar, despite the fact that he was the main deity of Larsa.[8] In most examples, he is shown in his characteristic ascending posture,

[5] Seen also on a fragment of an impression, *Ur Ex* 10, 539. *Nippur* I, Pl. 120, No. 12, the god standing over the platform holds a crook. Figulla, 1914, Pl. 14, Nos. 56, 57, show impressions from tablets dated to Warad-Sin; one depicts the god with the mace standing on a platform and possibly facing the goddess Ishtar. The figure with mace standing on a patterned platform at the end of an offering scene is illustrated on an Old Babylonian seal in Moorey, *Iraq* 35, Pl. XXXIV, No. 13. Notice that the garment of the suppliant goddess is similar to Porada, fig. 4.

[6] Mentioned in a number of texts as having been worshipped at Larsa in association with the goddess Nana'a; see Kingsbury, 34, pp. 1-34.

[7] The seals are either an imitation of the seals of the Sippar workshops, or imported from that center. In fact, traces of the characteristic vertical undulating lines on No. 277a and on the worshipper's garment on 277b, points to the fact that they were imported.

[8] Goetze, *JCS* 4, p. 90.

placing one foot on a platform, Pl. XXXIX: 1, 2 and 3, or on a kneeling human-faced bull. In Pl. XXXIX: 4, Shamash is enthroned resting his feet on a human-headed reclining bull. The enthroned god in Pl. XXXIX: 5 is probably also Shamash, considering his high-backed throne which is usually associated with the sun god (already discussed with examples from Harmal, Pl. I: 10). Unusual in Pl. XXXIX: 5 is the staff with a globe at each end, held by the god.

The filler motifs of Pl. XXXIX: 2 associate the seal with a group of impressions illustrating smaller motifs arranged in two registers as seen in examples from Ur, also dated to the Larsa dynasty (Pl. XLII: 7, 8 and 9. The sealing of Pl. XXXIX: 2 shows the metal caps, a detail which occurs on other impressions from Larsa, Pl. XXXIX: 6, and Porada, fig. 14.

3.1.3. Other Gods

The only other identifiable deity present on the impressions from Larsa is the god Adad seen on Porada, Fig. 14. In Pl. XXXIX: 7, he is standing on a lion-dragon, while on Pl. XXXIX: 8, he is portrayed wearing a long pleated garment, with one foot on a kneeling bull and holding a triple lightning fork.

The very fine seal of Puzur-Ninkarrak[9] depicts the god Nergal as a conquering god. Nergal or Ninurta are possibly portrayed in the impression of Pl. XXXIX: 9 in which a lion-headed demon holding a victim must have faced the god in the original seal. Unusual is the posture of the small figure of the warrior pointing an arrow upward.[10]

3.1.4. The Figure with Mace

Second to Ishtar and Shamash, the figure with the mace is most frequently represented. In the earlier impressions he seems to have played a minor role, normally placed at the end of a ritual scene, Pl. XXXIX: 3, or facing the goddess Ishtar, Pl. XXXVIII: 1. Later, particularly during the time of Rim-Sin (c. 1822-1763 B.C.), he is frequently depicted facing a suppliant goddess alone, Pl. XL: 1, 2 and 3.[11] In Pl. XL: 4, there are two columns of inscriptions between the figure with the mace and the goddess. Columns of inscription with the figure of the suppliant goddess facing them was a popular subject on the seals of the Larsa dynasty.[12]

It is noticeable that the figure with the mace accompanied by the suppliant goddess and an inscription becomes the main motif at Larsa during the reign of Samsuiluna, Pl. XLI. Pl. XLI: 2 depicts him facing the interceding goddess with the columns of inscription in between. The seal of Pl. XLI: 6 was originally capped and there is emphasis on exposed drill work. In Pl. XLI: 4, the figure is facing the double lion-headed club (top missing). Pl. XLI: 5 shows a small male figure in reverse position over a second male figure. The last feature was common in the repertoire of the later part of the Old Babylonian period.

[9] Porada and Basmachi, *Sumer* 7, p. 68.

[10] I was not able to draw the impression accurately due to a modern break.

[11] *VR*, 322; Figulla, 1914, Pl. IV, Figs. 83, 86, 92; Faust, *YOS* 8, Pl. VII, No. 12.

[12] *Brett*, 76; *Newell*, 661; Figulla, 1914, Pl. III, No. 66; Faust, *YOS* VIII, Pl. LI, No. 115; also on a clay sealing from Ur, *Ur Ex.* 10, 459.

In addition to Pl. XXXVIII: 3, 4, 5, and Pl. XL: 7 which, it has been suggested above, were imported from the workshops of Sippar, is an impression on a Larsa tablet dated to Warad-Sin. Depicting a snake goddess and an acrobat, it was published by Buchanan,[13] who suggested a northern origin for the motif. He supported his argument by comparison with other examples of Syrian or northern manufacture.[14] The principal subject may well have been of foreign derivation, but the seal was most probably made at Sippar. The majority of subsidiary motifs occur on seals characteristic of the Sippar workshops. Similar, too, is their arrangement of the figures. Even the static but balanced and precise placing of the main figures is in the Sippar style, and the deities and worshippers fall within the repertoire of the Sippar glyptic. Moreover, the snake goddess seems to have been known to the Sippar craftsmen, as she occurs as a filler motif in an impression from the time of Samsuiluna.[15] As in the rest of the tablets from the sites discussed, impressions of Old Assyrian seals occur. At Larsa, an example is Pl. XL: 8.

Although the number of impressions from Larsa is limited, it is possible to form a general view of the seal engraver's craft, which seems to have been of very high quality. Distinctive features not found elsewhere, such as Ishtar holding only a mace, seem to be a local peculiarity. The repertoire and style was already developed by the reign of Abisare[16] and continued to the reign of Rim-Sin, so far as one can judge from the small collection. Simplification of subject matter (figure with mace, suppliant goddess and inscription) and the deterioration in the standard of manufacturing become apparent during the reign of Samsuiluna. This may be attributed to the loss of Larsa's independence to Hammurabi and thus its diminution as a cultural center.

3.2. Ur

This group of impressions proved to be the most difficult to study, because they are mostly on small tablets which were sealed before they were inscribed so that the text is written over the impression.[17] Whereas at Larsa and the Diyala sites the impressions are on the envelopes of the tablets, at Ur only one seal was usually rolled and this was done several times all over the entire tablet. Hence, only one section of the scene was repeated. This was possibly due to the seal being worn on a chain round the neck and being used while remaining on the owner's neck.[18] On the majority of tablets only the inscription of the seal is impressed and this accounts for the small amount of material presented here. It also indicates that at Ur, as at Larsa, most of the seals were inscribed, whereas at Sippar and in the Diyala they were not.

Most of these tablets belong to the period when Ur was under the control of Larsa as they are dated by the years of the Larsa kings.[19] It might be due to this practical orientation that there

[13] Buchanan, *Iraq* 33, pp. 1-18, Pl. Ia.

[14] Ibid., Pl. Ie.

[15] Lyon, *JAOS* 27, p. 135, Pl. 1-11. Published also in *Cyl. Seals*, Pl. XXX a. As a filling motif the snake goddess also appears on *VR* 479, and Tell Harmal No. 60c.

[16] Porada and Basmachi, *Sumer* 7, p. 66. Unfortunately few seals could be attributed to Larsa to determine the existence of a second style (inferior), like the early Old Babylonian group from the Diyala.

[17] The texts have been published in *UET* 5. The impressions come from the tablets that have been allotted to the Iraq Museum

[18] This is only a suggestion.

[19] The exceptions are an impression on a tablet and two seals dated to King Bur-Sin of Isin, representing the short period he gained control of the city. *Ur Ex* 10, 445, 528, 540.

is a complete absence of the figure of Nana-Suen, the paramount god of Ur, so far as this group of texts is concerned. Instead, the most frequently represented god is Shamash. An impression on a tablet dated to one of the early kings of Larsa, Sin-iddinam,[20] depicts a worshipper pouring a libation in front of the Sun God. A similar subject is also seen on a seal from a Larsa grave[21] where the god holds the ring and staff. The Sun God is normally represented in this attitude: one foot raised on a symbolic mountain. The platform is sometimes decorated with vertical incisions, Pl. XLII: 1,[22] or chequered pattern, Pl. XLII: 2 and 3.

The god with the mace wearing a horned headdress is depicted facing the Sun God in Pl. XLII: 4 and 5. In Pl. XLII: 5 there are three drill holes between them. Drill holes also appear on a seal from the Larsa period at Ur.[23] The Sun God on the seal stands upright and not in his normal ascending posture. He also wears a flounced garment, appropriate to deities in less active positions. Another seal, also from Ur,[24] shows a similar god, but not holding a saw. Most probably the Sun God is intended here; the god with the mace is facing him. The rendering of the garment of Shamash is so similar to a third seal from Ur[25] and so distinctive from other sites that they could all have come from one source.[26]

The seals from Ur exhibit the orthodox repertoire of the Old Babylonian period, but a local tendency can be recognized. The worshipper always wears a long garment that envelops the whole body; two exceptions are Pl. XLII: 3 and 6. This characteristic feature is also confirmed by the extant seals which show the worshipper always wearing the wrapped-up garment.[27] The worshipper's attire may indicate a southern fashion in dress that prevailed at Ur.

A second feature frequent on the seals from Ur is the terminal that consists of several motifs, forming two registers.[28] In Pl. XLII: 7, from the time of Sumuilum, a kneeling male figure is seen behind a lion who is attacking a victim (not preserved), while the lower register shows a contest scene.[29] Pl. XLII: 8 is of a similar character, though the figures differ in style. The subject of Pl. XLIII: 9 is rare: two nude females standing side by side are placed above two male figures who face each other. Antithetical grouping of identical motifs forming terminals (seen above in Larsa Pl. XXXIX: 2) seems to be characteristic of the early Old Babylonian period.[30]

Seals of original designs pointing probably to personal orders or importation appear in the early part of the Old Babylonian period. The half-kneeling nude deity behind the worshipper in Pl.

[20] Ibid., 442.
[21] Ibid., 478.
[22] Other examples from *Ur Ex 10*, 478, 479. Vertical incisions are depicted on the platform of Porada No. 267b, and *Ur Ex 10*, 530. Such a decoration could be a southern feature.
[23] Ibid., 498.
[24] Ibid., 500.
[25] Ibid., 496.
[26] A similar figure of the sun god is depicted on a seal said to come from Surgul, *VR*, 384. It should be noted that when Shamash is shown standing upright, he is always wearing the flounced garment (Nos. 157b, 227.4).
[27] Except *Ur Ex 10*, 483.
[28] Such a motif also occurs at other sites, Porada No. 267a, Tell Ed-Der and Sippar; but at Ur the terminal occurs even more frequently, which suggests a continuous tradition.
[29] An Old Assyrian style is attested here.
[30] See *Cyl. Seals*, Pl. XXVIe; two heads of bull-men above another two; also *Sippar* No. 185.3.

XLIII: 1 is a figure which can only be found on Agade seals. Pl. XLIII: 2 from the reign of Sin-iqisham illustrates the mythological theme of the battle of the gods, again derived from the Agade period but in an Old Babylonian form.[31] Although seals with a similar motif are illustrated in several private collections,[32] all examples portray a powerful god vanquishing an enemy with a weapon. The victim is seen in all the examples bending sharply from the waist, with his arms tied behind his back. His crown is lost but he is wearing the vertically pleated long skirt, reserved for divinities. Thus the victim is presumably a deity. In one example, the victim has already lost his head, indicating that the weapon held by the god is a cutting instrument.[33] The similarity in the rendering in all the examples suggests that the subject represents a particular mythological episode.

Pl. XLIII: 3 depicts an unusual aspect of the goddess Ishtar. She is portrayed in profile standing on two lions and holding a bow[34] instead of the usual mace-club.

Pl. XLIII: 4, dated to the 26th year of Rim-Sin, is possibly a revival of an earlier style and most probably copied from such examples. The original seal had a metal cap.

Pl. XLIII: 5 is also unusual in Old Babylonian glyptic. The impression depicts the god Usmu, Ea's assistant, standing over a goat-fish as a nude bearded hero with curls carrying a flowing vase stands by, and a pig is between them. Similarly, this motif is possibly derived from Agade seals, which disappeared from the Mesopotamian glyptic only to appear quite frequently on the Kültepe sealings.[35] The pig is depicted on one Old Babylonian seal where a goat-fish also appears,[36] giving support to the symbolism of the pig as an attribute of Ea, as already represented on Agade seals.[37]

There is no evidence of seals that came from the Sippar workshops, as at Larsa, with one exception: a fragment of an impression showing a suppliant goddess (Pl. XLIII: 8).

The god Amurru appears on one tablet from the time of Samsuiluna, Pl. XLIII: 7. The terminal here, two winged demons, one in inverted position, is typical of the post-Hammurabi period[38] as is the rendering of the goddess' garment, Pl. XLIII: 8, where the engraver left a plain wedge in the center, while the rest is engraved with the flounced pattern.

[31] Legrain, No. 329; Newell, 157; Van Buren, 1959, No. 88.
[32] Buchanan, *Yale Library Gazette* 45, No. 9.
[33] Van Buren, 1959, 88. Otherwise, it looks more likely that the god is strangling his victim with a rope.
[34] Ishtar in profile, *Sippar* No. 247.5.
[35] Ozgüç, 1965, pp. 60-61.
[36] Ibid. Also *Louvre* II Pl. 80, No. 10 (A.354).
[37] Ibid., Pl. 73, No. 3 (A.160).
[38] See Porada, No. 292; *Sippar*, Nos. III and V.

4

CONCLUSIONS

When this study was undertaken, it was my hope that I would be able to draw up a chronological table of motifs and styles represented in the Old Babylonian seal impressions examined. However, this proved to be difficult, because given motifs did not often occur simultaneously at the different sites. Nevertheless, it is possible to discern, on the basis of frequency, certain trends in motifs and styles at each site which may have been characteristic of the region in question. As has been indicated in the foregoing study, the styles of the sites examined in each of the three regions coincide to a considerable extent.

A table showing geographical distribution would also be difficult because of the limited quantity of material (from only six sites) discussed here. It is possible, nevertheless, to describe motifs and styles, at least tentatively, in terms of dates and regions.

4.1. The Diyala Region

The devotional scene is the main subject of Old Babylonian seals followed in frequency by friezes of fighting animals and monsters. At Tell Harmal these friezes are more abundant in the earlier levels IV and III. In fact, the motif of two pairs of contestants (Harmal, Pl. I: 2), a subject derived from Agade seals, appears only in the early levels. The more complex fighting scene continues in use until the middle of the Old Babylonian period (Harmal, Pl. I: 6). It occurs at Sippar until the time of Samsuiluna (Sippar, Pl. XXXIII: 8).

The devotional scene accompanied by a panel of inscription (normally bearing the name of Shamash and Aia), placed above an animal, another feature of Agade composition also occurs in the early levels. A similar panel of inscription occurs in compositions with two pairs of contestants.[1] Because this feature is common to both types of scenes, it is possible to suggest that both were the product of the same center during the earlier part of the Old Babylonian period.

The seals from Harmal are clearly of true Babylonian and not of local craftsmanship, although their execution does not compare with the standard of the early seals from Akkad or Sumer. The varied styles and motifs appearing at Harmal merely indicate that the seals were imported from various centers of Babylonia.

At Dhiba'i the contemporary impressions illustrate two styles: seals like those from Harmal produced in Babylonia proper and a second group of inferior seals in the Diyala styles, similar

[1] *Corpus*, 347.

to those earlier found at Tell Asmar. These were the product of a workshop located either at Dhiba'i, or in a neighboring town.

They always depict a presentation scene to an enthroned king who wears a round cap. The worshipper wears a similar cap, and is either led by the interceding goddess or precedes her. Frequently two worshippers appear in one scene. The crescent, vase and ball-staff are dominant filler motifs. What is peculiar to these seals is the rudimentary representation of the attire of the worshipper, goddess and king. The single-horned crown of the goddess is sometimes reduced to just a scratch. Peculiar to the goddess are the crossed straps across her chest, better seen on the Asmar seals.

A certain number of Old Assyrian seal impressions occur on the Dhiba'i and Harmal tablets, illustrating the close connections between the Diyala and the north. This may also explain a number of motifs which occur later on Mitannian seals, e.g. the dancing men (Harmal, Pl. XIV: 8) or the winged goddess (Harmal, Pl. IX: 7). The crossed winged lion-demons superimposed on the nude hero (Harmal, Pl. I: 9) also have connections with the Old Assyrian style.[2]

There is little variety in the inscriptions on the early seals which are usually restricted to the name of Shamash, the Sun God. By the time of King Naram-Sin (ca. 1840 B.C.) of Eshnunna (Level III), seals inscribed with the name of the owner became more frequent. Inscribed seals are predominant during the reign of Ibalpiel II, a contemporary of Hammurabi (Level II, ca. 1780 B.C.). At this time there is a change in the style and repertoire of the seals, with a simplification of the scenes and an almost total disappearance of the introduction scene. Few gods remain and mythological scenes are nonexistent. Instead, in Level II, the majority of the impressions have only the figure with the mace and the suppliant goddess beside the inscription. Similar seals during this period occur at Sippar but alongside seals portraying the devotional scenes of Workshops I and II. They appear later during the reign of Samsuiluna at Larsa. Gold-capped seals decorated in both filigree and granulation occurred at Harmal (Level II, Pl. XI: 5, 9, 10 and 11) earlier than those from Sippar (Pl. XXXVI: 1), where they make their appearance in the post-Hammurabi era.

4.2. Akkad

From the large number of impressions from Sippar and Tell Ed-Der, a few points have been established. First and foremost are the high quality of the engraving the the variety of subjects used by the seal cutters of the early period. Every single impression of the envelope of Sippar No. 182 (Pl. XXXVIII) confirms the excellent engraving of the original seals. It seems the craft was already well established here by continuing the Agade and Ur III glyptic traditions. There is no sign of inferior workmanship. Mythological and contest friezes were also popular, Pl. XXII; 5, Pl. XXIII: 1, and Pl. XXIV: 1.

Two distinct local workshops emerged during the reign of King Sin-muballiṭ to flourish during the time of Hammurabi and Samsuiluna. Both produced the best works of the Old Babylonian period. The main subject they illustrate is the devotional scene, with the principal Mesopotamian gods represented. A worshipper is usually seen offering a kid to the deity. The main god, when enthroned on a panelled stool, always holds the staff or ring and a staff. In the Diyala (Harmal)

[2] *VR*, 513; Buchanan, *Yale Library Gazette* 30, No. 57.

and in Sumer, contemporary seals did not depict enthroned deities which was a motif peculiar to Sippar. The assembly of several gods on one seal is also characteristic of the Sippar seals, but it is difficult to identify the function of the variety of minor deities depicted on these seals. A goddess (Sippar, Pl. XXVII: 6 and 7) with one hand raised seems to be performing the act of introduction. A god occurs wearing a plain garment like a worshipper with his hands folded on his chest (Sippar, Pl. XXVIII: 5), and a mortal wearing a cap with upturned brim is depicted in the flounced garment of the gods (Sippar, Pl. XXVIII: 4) and offers a kid to a deity. A god offering a kid to the mian deity (Sippar, Pl. XXVIII: 6) is unusual but not unique.[3] Various types of crowns are worn by minor deities, most curious of which is the square-topped horned crown. Both the nude attendant with three pairs of curls and bull-man also present kids. Ea's nude attendant who normally holds the flowing vases to his chest is seen on the Sippar impressions with the vase held away from his body. The filler motifs are drawn from the usual Old Babylonian repertoire. What is distinctive about the Sippar seals is the crowding of the seal with a number of standing figures interspersed with small motifs. The seals are easily recognizable in museum collections.

It is the technique of the cutting of these seals which indicates the possible existence of two workshops. Workshop I has, as its most distinctive feature, the rendering of the flounced garment of the gods, which consists of rows of triple undulating lines. A second characteristic is the practice of scarcely allocating space for the inscription, particularly notable in the Hammurabi and Samsuiluna eras when inscribed seals were popular in Sumer and the Diyala region. This occurred frequently among the seals from Workshop II. Instead, the signs of the inscription are scattered in the field or placed in a box as if the inscription had been added afterward.

The seal cutter of Workshop II carried on the tradition of the typical Old Babylonian styles in rendering his figures. The flounces have straight vertical lines. Columns of inscriptions are more frequent, and a large number of these seals depict only the god or the figure with mace facing a suppliant goddess.

By the end of the reign of Hammurabi, and more often during that of his successor, Samsuiluna, there is a noticeable change in the technique of cutting the seals. An increase in the use of the drill is apparent. At first drill holes were left unworked, probably for decorative reasons, and were used for such features as the fringed border of the worshipper's garment (Sippar, Pl. XXIX: 6), or the chignon of the god. The drill hole is more obvious in the beard of the god in Pl. XXXIV: 1, and in the arms and legs of the figure with the mace on Pl. XXXIV: 5.

In fact, this technique marks the beginning of the gradual change in glyptic art during the post-Hammurabi era. Not only did the drill dominate the rendering of the subjects, but the seals became smaller in size, particularly during the reigns of Ammiditana and Ammiṣaduqa. This was accompanied by a deterioration in the execution and a limitation in the actual subjects, with the filler motifs virtually disappearing. The horned crown was replaced by a pointed cap surrounded by drill holes (Sippar, Pl. XXXV: 6), or stylized further by a horizontal stroke topped by a large drilled circle (Sippar, Pl. XXXVII: 1 and 4). This exaggerated use of the drill resulted in the distortion of the figures (Sippar, Pl. XXXVII: 2).

[3] *Bib. Nat.*, 170.

After the death of Hammurabi, his successor Samsuiluna had to fight insurrections in Babylonia and the provinces. In his ninth year he encountered a "Kassite host" which was presumably defeated. The Kassite threat to Babylon remained in the background up to the end of the dynasty. In his twenty-eighth year Samsuiluna lost control of much of Sumer to the rising dynasty of the Sealand. He also fought against rebels in the northeast (Eshnunna) and the Upper Euphrates. Abi-eshuh's reign was also unsettled, and during his time Babylon lost control of the Middle Euphrates (Khana).[4]

The letters and administrative documents from the reign of Ammiṣaduqa indicate that his control of the countryside was limited as tribes were plundering the land. The gravity of the situation is illustrated by letters concerning instructions to close the gates of Sippar at night.[5]

This political weakness of the dynasty must have resulted in an unsettled economic situation, leading possibly to more expensive labor. This may explain the unskilled work in the cutting of the seals, with the use of the drill becoming dominant.[6] Seals are smaller in size and may have been cut from older seals. Hematite, the main stone used for making Old Babylonian seals, could have become scarce as a result of the obstruction of trade from its place of origin. Yet it is difficult to suggest how the style developed at Sippar, owing to the scarcity of comparative material from the surrounding areas. It is found in the west, as drilled-type Old Babylonian seals contemporary with Abi-eshuh and later have been found in impressions on the Alalakh tablets of Level VII.[7]

Relations between Sippar and the north and west are illustrated by a number of Old Assyrian (Pl. XLIV: 1-9) and Old Syrian (Der and Sippar, Pl. XXVII: 11 and 12) impressions on tablets from the *nadītu* archive at Sippar and from Tell Ed-Der. Sippar-type seals from both Workshops I and II occur at Mari.[8] The style was also imitated in Syria; this is apparent in the engraving of the suppliant goddess, or the enthroned god wearing the flounced garment with triple undulating lines.[9]

4.3. Sumer

Sumer in the early Old Babylonian period gives a different picture from the Diyala. A uniform style is apparent in the impressions from both Larsa and Ur, due most probably to the political domination of Larsa over southern Babylonia. The seals display skilled craftsmanship, and the Old Babylonian style appears as early as King Gungunum of Larsa (Ur, Pl. XLIII: 6): the suppliant goddess is engraved with all the careful detail that is to be seen on later seals of the Hammurabi period from Sippar.

[4] Gadd, *Cambridge Ancient History* Vol. II, Part I, pp. 220-224.

[5] Leemans, *JESHO* 11, p. 175.

[6] This is the beginning of the mechanization of seal engraving which reaches its final form in the 15th century B.C. with the seals made in the kingdom of Mitanni, cut entirely by drilling. An earlier example of this technique is a class of seals dating to the Jamdat Nasr period. In both cases the explanation was probably a shortage of skilled craftsmen and the large demand for seals. The Mitanni had fine seal engravings as we know from the princely seals. However, this explanation does not answer the problem of the Sippar seals.

[7] Collon, 1975, Nos. 128, 129, 132, 135, p. 143.

[8] *MAM II*, Pl. LIII, No. 246; Pl. LV, No. 285, for Workshop I; and Pl. LVI, No. 32 for Workshop II.

[9] Ibid., Pl. XLVII, No. 71a; *Corpus*, 910.

Like the early seals from the Diyala and Akkad, the influences of Agade glyptic is strong, as is well illustrated by the seal of Puzur-Ninkarrak. Mythological scenes were also popular. Pl. XLIII: 2, from Ur, shows a battle among the gods, a subject derived from Agade seals. Antithetical grouping of identical motifs forming terminals occur on the impressions from the early Old Babylonian period (Der, Pl. XXIII: 5; Sippar, Pl. XXIII: 6; Larsa, Pl. XXIX: 2; and Ur, Pl. XLII: 9). The complex fighting friezes in two registers (Nurshamash archive XXXXXX; Der, Pl. XXIII: 3; Sippar, Pl. XXII: 4; and Ur, Pl. XLII: 7) can also be assigned to the early Old Babylonian period.

The impressions published by Porada from the reign of King Sin-iddinam are another example of the high standard of seal engraving in southern Babylonia during this period. The majority of seals are inscribed, in contrast to the rare occurrence of inscriptions on contemporary impressions from Harmal and Dhiba'i (except for the class of seals which prevailed at Harmal during the time of Ibalpiel II). At Larsa, similarly inscribed seals with the figure with the mace and the suppliant goddess occur more frequently on tablets dated to King Samsuiluna, thus postdating the examples at Harmal.

Characteristic of the early seals at Larsa is the special care given to detail, for example Ishtar's costume in Pl. XXXVIII: 1. Her lion is also carefully executed, again in contrast to the stylization used on the Harmal seals. In addition there is a local motif seen uniquely on the impressions from Larsa showing the goddess Ishtar holding only a mace (Larsa, Pl. XXXVIII: 6).

4.4. Use of Seals

The use of the seal on written documents during the Old Babylonian period differed from one region to another and varied according to date. At Harmal in Levels IV and III, the impressions are found on the envelopes, and this applies also to Dhiba'i. The practice change in Level II at Harmal where the tablets are small in size and many are without envelopes. The impressions on the tablets are fewer, usually reduced to one seal rolled with the emphasis on the inscribed part of the seal all over the tablet before the writing of the text. Thus, much of the seal impression is obliterated by the cuneiform signs of the text.

At Ur, the tablets were sealed in the same fashion throughout the Old Babylonian period, and the same applies to the tablets at Larsa. Among the latter, a few impressions were made on both envelopes and tablets, particularly during the reign of Warad-Sin.

At Sippar, up to the time of Samsuiluna, the seals were rolled on the envelopes. Noticeable are the large number of seals used on each document. A change appeared in the reigns of Ammiditana and Ammiṣaduqa similar to the change mentioned at Harmal during the time of Ibalpiel II: tablets were not always encased in envelopes and they normally had no more than two impressions, mostly damaged by the text of the document.

The partial obliteration of the impression by the cuneiform text and the incomplete rolling of the seals are two factors that limit the study of the Old Babylonian sealings. The number of sealings on a document also varied from one site to another, depending on the period. At Harmal, in Levels IV and III, up to five impressions are found on the envelopes, while in Level II one impression was the rule. At Sippar some envelopes had more than ten sealings. A reduction in the number of impressions to one or two accompanied the change to dispensing with an envelope and to sealing the tablet only. This occurred after the reign of Samsuiluna when the drilled style replaced the earlier, carefully modelled, one on Old Babylonian cylinders. In Sumer (Ur and Larsa) few seals are rolled, not more than two or three. The practice was the same throughout the period. These variations in the number of persons required to seal a document could be the result of administrative requirements. Although there is a variety in the size and shape of the tablets, the majority of the texts are economic documents, which required sealing. These comprise loans, leases, sales, documents concerning inheritance, and receipts. There are also a few examples of sealed letters.

4.5. Final Remarks

There is no uniformity of style among the seals of the Old Babylonian period. Nor can the style and choice of motif be used uniformly as a chronological index. This can only be done in relation to individual sites or areas. With the limited material at our disposal it has been possible to identify a few regional styles and place them in their chronological setting. There is sufficient indication to suggest the occurrence of more local workshops with distinctive styles in other major cities. Our data and the approach suggested here may, hopefully, contribute to the understanding of further practices as evidence becomes available.

CHRONOLOGICAL TABLE

BABYLON	LARSA	ESHNUNNA	HARMAL
	Abisare 1905-1895		
Sumuabum 1894-1881		Abdi-Erah	Level IV Sin-abushu
		Belakum	
			Hammidashur
Sumulael 1880-1845			
		Warassa	
			Level III Hammidashur
		Ibalpiel I	
	Sin-iddinam 1849-1843		
Sabium 1844-1831			
		Naram-Sin	Naram-Sin
	Warad-Sin 1834-1823		
Apil-Sin 1830-1813			
Sin-muballiṭ 1812-1793	Rim-Sin 1822-1763		Level II
		Dadusha	
Hammurabi 1792-1750			
		Ibalpiel II	Ibalpiel II
Samsuiluna 1749-1712			
Abi-eshuh 1711-1684			
Ammiditana 1683-1647			
Ammiṣaduqa 1646-1626			
Samsuditana 1625-1595			

ABBREVIATIONS

AASOR	Annual of the American Schools of Oriental Research.
AfO	Archiv für Orientforschung.
AJSL	American Journal of Semitic Languages and Literatures.
Al-Hashimi	Ridha al-Hashimi, "Some Old Babylonian Purchase Contracts in the Iraq Museum." M.A. dissertation, University of Baghdad, Iraq, 1964.
AOAT	Alter Orient und Altes Testament.
Ashmolean	Briggs Buchanan, *Catalogue of Ancient Near Eastern Seals in the Ashmolean Museum, I: Cylinder Seals*. Oxford, 1966.
Baqir	Taha Baqir, "Date-Formulae and Date-Lists from Harmal." *Sumer* 5, 1949.
BE	The Babylonian Expedition of the University of Pennsylvania.
Bib. Nat.	Louis Delaporte, *Catalogue des cylindres orientaux et des cachets de la Bibliothèque Nationale*. Paris, 1910.
Boehmer	R. M. Boehmer, *Die Entwicklung der Glyptik während der Akkad-Zeit.* (Untersuchungen zur Assyriologie und vorderasiatischen Archäologie, Bd. 4) Berlin, 1965.
Brett	H. H. Von der Osten, *Ancient Oriental Seals in the Collection of Mrs. Agnes Baldwin Brett*. (OIP XXXVII) Chicago, 1936.
Corpus	Edith Porada and Briggs Buchanan, *Corpus of Ancient Near Eastern Seals in North American Collections, I: The Collection of the Pierpont Morgan Library*, (The Bollingen Series XIV) Washington D.C., 1948.
CT	H. H. Figulla, *Cuneiform Texts from Babylonian Tablets in the British Museum, XLVII: Old Babylonian Nadītu Records*. London, 1967.
Cyl. Seals	Henri Frankfort, *Cylinder Seals, A Documentary Essay on the Art and Religion of the Ancient Near East*. London, 1939.
Danish Ntl. Museum	Otto Ravn, *A Catalogue of Oriental Cylinder Seals and Seal Impressions in the Danish National Museum*. Copenhagen, 1960.
DF	Date Formula.
Dh.	Tell Dhiba'i.
Diyala	Henri Frankfort, *Stratified Cylinder Seals from the Diyala Region*. (OIP LXXII) Chicago, 1955.
Edzard	D. O. Edzard, *Altbabylonische Rechts- und Wirtschaftsurkunden aus Tell ed-Der im Iraq Museum, Baghdad*. München, 1970.
Gordon	C. H. Gordon, "Western Asiatic Seals in the Walters Art Gallery." *Iraq* 6, 1939.
Harris	Rivkah Harris, *Ancient Sippar: A Demographic Study of an Old-Babylonian City (1894-1595 B.C.)*. Leiden, 1975.

HL	Tell Harmal.
HUCA	*Hebrew Union College Annual.*
IM	Iraq Museum.
IRSA	E. Sollberger and J. R. Kupper, *Inscriptions royal sumériennes et akkadiennes.* Paris, 1971.
JCS	*Journal of Cuneiform Studies.*
Jean	C. F. Jean, *Textes cunéiformes du Louvre, X & XI: Contrats de Larsa.* Paris, 1926.
JNES	*Journal of Near Eastern Studies.*
JESHO	*Journal of the Economic and Social History of the Orient.*
Karim	Abdul-Karim Abdullah Ahmed, "Old Babylonian Loan Contracts in the Iraq Museum from Tell al-Dhiba'i and Tell Harmal." M.A. dissertation, University of Baghdad, Iraq, 1964.
Legrain	Leon Legrain, *The Culture of the Babylonians from their Seals in the Collections of the Museum.* (*PBS* XIV) Philadelphia, 1925.
Louvre	Louis Delaporte, *Catalogue des cylindres, cachets et pierres gravées de style oriental. I: Fouilles et missions 1920; II: Acquisitions 1923.* Musée National du Louvre, Paris.
MAM II	André Parrot, *Mission archéologique de Mari, II: le palais.* Paris, 1959.
Moore	A. G. Eisen, *Ancient Oriental Cylinders and other Seals with a Description of the Collection of Mrs. William H. Moore* (OIP XLVII) Chicago, 1940.
Newell	H. H. von der Osten, *Ancient Oriental Seals in the Collection of Mr. Edward T. Newell.* (OIP XXII) Chicago, 1934.
Nippur I	D. E. McCown, R. C. Haines, and D. P. Hansen, *Nippur I, Temple of Enlil, Scribal Quarter and Surroundings.* (OIP LXXVIII) Chicago, 1967.
OIP	Oriental Institute Publications, The University of Chicago.
Porada	Edith Porada, review of *Corpus. JCS* 4, 1950.
PBS	Publications of the Babylonian Section, University Museum, University of Pennsylvania.
RA	Revue d'assyriologie et d'archéologie orientale.
Raschid	Fawzi Raschid, "Archiv des Nurshamash und andere Derlehensurkunden aus der altbabylonischen Zeit." Inaugural Dissertation, Ruprecht Karl Universität. Heidelberg, 1965.
RlA	*Reallexikon der Assyriologie und vorderasiatischen Archäologie.*
Sippar	Edith Porada, "The Seal Impressions," in Mogens Weitemeyer, *Some Aspects of the Hiring of Workers in the Sippar Region, at the Time of Hammurabi.* Copenhagen, 1962.
Suleiman	Amer Suleiman, "A Study of Land Tenure in the Old Babylonian Period with Special Reference to the Diyala Region, Based on Published and Unpublished Texts." Ph.D. Dissertation, School of Oriental and African Studies. London, 1966.
Susa	Pierre Amiet, *Mémoires de la délégation archéologique en Iran, XLIII: glyptique susienne.* Paris, 1972.
ThD	François Thureau-Dangin, *Lettres et contrats, de l'époque de la première dynastie babylonienne.* Paris, 1910.

UrEx10	Leon Legrain, *Cylinders. UrEx 10,* London, 1951.
UET 5	H. H. Figulla and W. J. Martin, *Letters and Documents of the Old Babylonian Period. UET 5,* London, 1953.
VR	Anton Moortgat, *Vorderasiatische Rollsiegel: ein Beitrag zur Geschichte der Steinschneide Kunst.* Berlin, 1940.
Yale	Briggs Buchanan, *Early Near Eastern Seals in the Yale Babylonian Collection.* New Haven, 1981.
ZA	*Zeitschrift für Assyriologie und vorderasiatische Archäologie.*

5

APPENDIX A

List of the Seal Impressions: Museum Numbers, Catalog Numbers, and Dates

5.1 Tell Harmal

IRAQ MUSEUM NO.	CATALOG NO.	DATE	YEAR
51057	97	Ibalpiel II	
51177	73	Ibalpiel II	7
51179	91	Ibalpiel II	
51190	87	Ibalpiel II	
51195	74	Ibalpiel II	8
51203	95	Ibalpiel II	
51230	113		
51251	83		
51289	85	Ibalpiel II	
51296	80	Ibalpiel II	
51331	70	Dadusha	
51349	99		
51468	75	Ibalpiel II	8
51541	90	Ibalpiel II	
51542	94	Ibalpiel II	
51609	61		
51622	114		
51657	89	Ibalpiel II	
51658	86	Ibalpiel II	
51741	78	Ibalpiel II	
51788	82	Ibalpiel II	
52138	111		
52139	110		
52156	88	Ibalpiel II	
52178	79	Ibalpiel II	
52188	96	Ibalpiel II	
52268	108		
52275	81	Ibalpiel II	
52416	105		
52599	76	Ibalpiel II	10
52601	72	Ibalpiel II	5
52623	54	Naram-Sin	
52626	77	Ibalpiel II	10

IRAQ MUSEUM NO.	CATALOG NO.	DATE	YEAR
52688	67		
52889	106		
52913	100		
52937	107		
52944	112		
52950	109		
52986	102		
54029	103		
54340	101		
54446	71	Ibalpiel II	1
54684	47		
54686	42	Hammidashur	
54695	44	Hammidashur	
54983	48	Naram-Sin	
54985	46	Hammidashur	
54991	62		
55116	5		
55148	20	Sin-abushu	
55153	21	Sin-abushu	
55155	43	Hammidashur	
55158	22	Sin-abushu	
55161	45	Hammidashur	
55374	14	Sin-abushu	
55376	13	Sin-abushu	
55380	15	Sin-abushu	
55382	8	Sin-abushu	
55386	12	Sin-abushu	
55398	37	Ibalpiel I	
55433	104		
55457	31		
55567	34		
63125	33		
63134	11	Sin-abushu	

IRAQ MUSEUM NO.	CATALOG NO.	DATE	YEAR
63135	60		
63139	19	Sin-abushu	
63161	41	Hammidashur	
63163	63	Hammidashur	
63168	39		
63171	35	Hammidashur	
63172	38	Hammidashur	
63177	40	Hammidashur	
63185	25		
63187	24		
63192	57		
63197	27		
63203	58		
63213	49	Naram-Sin	
63214	65		
63218	23		
63232	9	Sin-abushu	
63236	17	Sin-abushu	
63237	18	Sin-abushu	
63240	16	Sin-abushu	
63241	1		
63242	3		
63243	4		

IRAQ MUSEUM NO.	CATALOG NO.	DATE	YEAR
63247	28		
63248	32		
63249	30		
63251	56	Naram-Sin	
63252	50	Naram-Sin	
63253	55	Naram-Sin	
63254	51	Naram-Sin	
63255	52		
63256	53		
63268	64		
63278	6		
63279	7		
63295	29		
63299	10	Sin-abushu	
63314	59		
63315	2		
67009	68		
67340	98		
H1.3-21	115		
H1.5-31	66		
H1.7-114	69		
H1.7-26	26		
H1.7-350	63		

5.2 Tell Dhiba'i

IRAQ MUSEUM NO.	CATALOG NO.	DATE DATE	YEAR YEAR
52770	120		
52772	118		
52782	117		
52787	116	Hammidashur	
52853	122		
52872	123		
54045	129		
66978	119		

IRAQ MUSEUM NO.	CATALOG NO.	DATE	YEAR
66981	125		
67273	124		
Dh.2-440	128		
Dh.2-441	130		
Dh.2-450	131		
Dh.2-571	127		
Dh.2-574	126		

5.3 The Archive of Nurshamash

IRAQ MUSEUM NO.	CATALOG NO.	DATE	YEAR
43552	1		
43555	2		
43556	3		
43559	4		
43560	5		
43562	6	Sin-abushu	
43591	7		
43654	8	Sin-abushu	
43665	9	Sin-abushu	
43666	10	Sin-abushu	
43668	11	Sin-abushu	
43684	12		
43688	13	Sin-abushu	
43818	14	Sin-abushu	
43819	15		
43827	16		
43843	17		
43882	18	Sin-abushu	
43885	19		
43886	20	Sin-abushu	
43897	21		
43908	22		

5.4 Tell Ed-Der

IRAQ MUSEUM NO.	CATALOG NO.	DATE	YEAR
48822	5		
48826	7		
49254	3		
49265	9		
49301	2		
49320	8		
49347	14		
49541	11		
49959	4		
50413	6		
50414	23	Ammiṣaduqa	4
50418	24	Ammiṣaduqa	7
50419	27	Ammiṣaduqa	14
50421	25	Ammiṣaduqa	7
50426	13		
50427	10		
50432	28	Ammiṣaduqa	
50437	15		
50440	17		
50471	16		
50498	18	Samsuiluna	18
50868	19	Ammiṣaduqa	2
50869	26	Ammiṣaduqa	11
50892	22	Ammiṣaduqa	4
50893	21	Ammiṣaduqa	4
50894	20	Ammiṣaduqa	4
50898	12		
51024	1	Sabium	13

5.5 Sippar

BRITISH MUSEUM NO.	CATALOG NO.	DATE	YEAR
16815	209	Hammurabi	14
16819	190	Sin-muballiṭ	
16821	213	Hammurabi	25
16823	188	Sin-muballiṭ	9
16824	203	Hammurabi	7
16826	185	Apil-Sin	
16827	202	Hammurabi	7
16833	191	Sin-muballiṭ	

BRITISH MUSEUM NO.	CATALOG NO.	DATE	YEAR
16834	223	Samsuiluna	1
16835	208	Hammurabi	11
16836	192	Sin-muballiṭ	
16841	200	Hammurabi	3
16851	240	Samsuiluna	14
16858	193	Sin-muballiṭ	
16861	246	Samsuiluna	
16862	229	Samsuiluna	6
17439	236	Samsuiluna	9
17045	205	Hammurabi	10
17046	207	Hammurabi	11
17049	247	Samsuiluna	
17051	235	Samsuiluna	9
17052	194	Sin-muballiṭ	
17054	224	Samsuiluna	1
17055	226	Samsuiluna	4
17056	201	Hammurabi	4

BRITISH MUSEUM NO.	CATALOG NO.	DATE	YEAR
17060	186	Apil-Sin	
17062	210	Hammurabi	
17063	221	Hammurabi	
17064	215	Hammurabi	32
17069	212	Hammurabi	25
17070	187	Sin-muballiṭ	8
17071	211	Hammurabi	14
17072	222	Hammurabi	
17085	227	Samsuiluna	4
17093	220	Hammurabi	
17099	184	Apil-Sin	
17194	219	Hammurabi	
78258	245	Samsuiluna	
78280	204	Hammurabi	
80161	261	Ammiṣaduqa	17
80161	251	Ammiditana	6
80217	256	Ammiditana	
82050	182	Buntahtunila	

LOUVRE MUSEUM NO.	CATALOG NO.	DATE	YEAR
AO. 1646	189	Sin-muballiṭ	
1647	195	Sin-muballiṭ	
1648	216	Hammurabi	35
1649	198	Sin-muballiṭ	
1650	197	Sin-muballiṭ	
1651	206	Hammurabi	10
1656	231	Samsuiluna	6
1657	233	Samsuiluna	8
1658	228	Samsuiluna	4
1671	252	Ammiditana	15
1672	199	Hammurabi	2
1677	230	Samsuiluna	6
1678	242	Samsuiluna	18
1682	225	Samsuiluna	2
1684	234	Samsuiluna	8
1726	241	Samsuiluna	14
1727	249	Abi-eshuh	
1733	217	Hammurabi	35
1734	257	Ammiṣaduqa	8

LOUVRE MUSEUM NO.	CATALOG NO.	DATE	YEAR
AO. 1750	232	Samsuiluna	8
1763	196	Sin-muballiṭ	
1766	183	Sumulael	
1924	243	Samsuiluma	22
1928	214	Hammurabi	26
1931	260	Ammiṣaduqa	16
2500	258	Ammiṣaduqa	8
2502	250	Ammiditana	5
2503	254	Ammiditana	34
2708	238	Samsuiluna	10
2709	253	Ammiditana	24
2713	239	Samsuiluna	11
2720	228	Samsuiluna	4
4139	237	Samsuiluna	10
4498	244	Samsuiluna	33
4499	255	Ammiditana	37
4649	248	Samsuiluna	
7595	257	Ammiṣaduqa	8

5.6 Larsa

LOUVRE MUSEUM NO.	CATALOG NO.	DATE	YEAR
AO. 6346	1	Sin-iqisham	
6347	2	Sin-iqisham	
6348	14	Rim-Sin	2
6349	13	Rim-Sin	2
6350	5	Warad-Sin	9
6351	6	Warad-Sin	9
6352	8	Warad-Sin	11
6353	7	Warad-Sin	10
6355	4	Warad-Sin	9
6356	9	Warad-Sin	12
6357	10	Warad-Sin	10
6358	21	Rim-Sin	17
6359	23	Rim-Sin	23
6360	24	Rim-Sin	25
6362	25	Rim-Sin	36
6364	18	Rim-Sin	10
6369	16	Rim-Sin	8

LOUVRE MUSEUM NO.	CATALOG NO.	DATE	YEAR
AO. 6371	17	Rim-Sin	9
6374	20	Rim-Sin	15
6375	19	Rim-Sin	10
6376	11	Warad-Sin	
6377	15	Rim-Sin	8
6380	3	Warad-Sin	9
6395	22	Rim-Sin	18
6396	33	Samsuiluna	
6405	28	Samsuiluna	6
6409	29	Samsuiluna	7
6414	26	Samsuiluna	1
6422	30	Samsuiluna	7
6423	27	Samsuiluna	3
6427	32	Samsuiluna	
6428	12	Rim-Sin	2
6440	31	Samsuiluna	7

5.7 Ur

IRAQ MUSEUM NO.	CATALOG NO.	DATE	YEAR
57206	12	Rim-Sin	
57221	10	Warad-Sin	11
57225	3	Simu-ilum	4
57233	14	Rim-Sin	1
57235	26	Rim-Sin	31
57264	2	Sumu-ilum	3
57271	23	Rim-Sin	26
57292	25	Rim-Sin	29
57300	31	Rim-Sin	35
57304	32	Samsuiluna	23
57307	17	Rim-Sin	7
57318	5	Sin-iqisham	5
57319	8	Warad-Sin	4
57320	11	Warad-Sin	11
57334	13	Rim-Sin	1
57337	9	Warad-Sin	10

IRAQ MUSEUM NO.	CATALOG NO.	DATE	YEAR
57345	18	Rim-Sin	10
57358	21	Rim-Sin	15
57359	22	Rim-Sin	17
57375	6	Warad-Sin	3
57376	19	Rim-Sin	10
57379	15	Rim-Sin	2
57382	7	Warad-Sin	3
57383	16	Rim-Sin	16
57385	24	Rim-Sin	26
57389	30	Rim-Sin	34
57390	29	Rim-Sin	34
57408	1	Gungunum	19
57449	20	Rim-Sin	11
57589	27	Rim-Sin	27
57594	28	Rim-Sin	33
57611	4	Sumu-ilum	9

6

APPENDIX B

Catalogue of the Seal Impressions *

6.1. Tell Harmal

6.1.1. LEVEL IV

1.

a. Part of contest scene. Bull-man fighting with lion. 21 mm. (Pl. I-1)

b. Worshipper wearing long garment open in front and a cap with upturned brim, offering kid to enthroned deity holding the ring and rod (crown and throne are missing). 20 mm. (Pl. IV-1)
HL. 7-225; Rm. 526; IM 63241.
Envelope; adoption document, *Karim* 40.

2.

a. Worshipper wearing long, fringed garment and round cap led by interceding goddess towards an enthroned deity on platform (only part is visible). Crescent in sky. Inscription. 15 mm. (Pl. III-7)

b. Suppliant goddess behind worshipper offering kid, facing Shamash holding saw and stepping on artificial mountain. In sky: sun disc in crescent. Figure with mace (damaged) facing goddess Ishtar in profile, possibly holding double club-mace, bow slung over shoulder. 20 mm. (Pl. IV-7)
HL. 7-358; Rm. 509; L. III (heirloom); IM 63315.
Envelope; purchase of field; *Al-Hashimi* 28. (Textual evidence indicates that it comes from L. IV.)

3.

a. Nude bearded hero with three pairs of curls overpowering lion by holding it in reversed position. Suppliant goddess behind worshipper wearing cap with upturned brim, facing god on throne with curved back. In the field: sun standard, scorpion, vase, drill hole. 15 mm. (Pl. I-10)

b. Contest scene. Three lions attacking human victim, bull and goat, respectively. 12 mm. (Pl. I-3)

c. Worshipper with clasped hands, facing enthroned deity resting feet on crouching animal, probably goat or gazelle. Three columns of faint inscription. Suppliant goddess, outline of seated dog. In sky: sun disc in crescent. 19 mm. (Pl. III-5)
HL. 7-226; Rm. 589; IM 63242.
Envelope; purchase of field; *Al-Hashimi* 21. The text has an oath formula mentioning the name of Immerum, King of Sippar.

4.

a. Roughly-cut seal. Worshipper with one hand raised, led by interceding goddess towards enthroned figure probably holding cup. In the field: crescent above crook. Crescent standard with tassels. 13 mm. (Pl. XXI-1)

b. Better-cut seal, though faintly impressed, resulting in absence of many details. Suppliant goddess behind worshipper with hands folded on his chest, wearing cap with upturned brim and long garment. He faces an enthroned "king," also wearing round cap and holding cup. Behind latter stands figure with mace, before whom miniature lion seems to be standing on hind legs. In field: above, globe, eight-pointed star, sun disc in crescent; and below, two more drill holes. Terminal: above, head of bull-man; below, double lion-headed mace. 20 mm. (Pl. II-4)

c. Two pairs of contestants. Human-headed bulls attacked by bull-men. In field: double lion club and ball-staff. Terminal: crescent above reclining bull, separated by horizontal line from seated dog. 15 mm. (Pl. I-2)

d. Combat scene with figures' heads missing: nude hero overpowers bull; two lions attack goat. In field: mongoose and ball-staff. 13 mm. (Pl. I-7)
HL. 7-227; Rm. 589; IM 63243.
Envelope; adloption document; *Suleiman* 86.

*Catalogue numbers appear below seal impressions on Plates I-XLIV. Millimeter measurements refer to height. L. = level. Rm. = Room.

5.
Offering scene. Suppliant goddess stands behind worshipper wearing long garment open down the front. Worshipper offers kid to Sun God, who is in ascending posture and holds saw. Behind him, nude figure raising curved object. In sky: star disc in crescent. 20 mm. (Pl. V-7)
HL. 5-98; Rm. 361; IM 55116.
Envelope; purchase of field; *Al-Hashimi* 18.

6.
Suppliant goddess following bearded worshipper who wears long fringed garment and thick hair indicated by vertical parallel lines. This can only be seen in outline. Worshipper raises one hand, while other is held to chest; he faces enthroned "king" who wears similar hair style and holds staff. In field: star above vase, sun disc in crescent above ball-staff. Terminal: miniature scene of worshipper raising both hands before enthroned "king" holding cup. Below, two columns of inscription. 20 mm. (Pl. II-1)
HL. 7-429; IM 63278.
Envelope; loan contract; DF is similar to *Harris* No. 16 in *JCS* 9. 1955.

7.
Part of combat scene. Lion-griffin held by bull-man. 16 mm.
HL. 7-184, Rm. 527; IM 63279.
Envelope; sale of slave.

8.
a. Nude male figure holding curved weapon behind worshipper wearing long garment and round cap marked with vertical lines. 15 mm.
b. Back of deity turned to left, is followed by worshipper who has both hands clasped and held to waist, and wears long plain garment and round cap. In second group, suppliant goddess facing god with mace. In field: crescent above mongoose. 18 mm. (Pl. IX-1)
HL. 6-149; Rm. 447; IM 55382.
Envelope; unpublished DF assigned accordingly to Iraq Museum inventory, to Sin-abushu.

9.
Three fragments of one impression. Worshipper offering kid to god holding saw, seated on throne with low carved back decorated with circles. He rests feet on reclining human-faced bull. Nude priest with pail and sprinkler standing on niched platform, faces bull-eared god, full-face, wearing multiple crown and long garment that opens down front. He holds one hand to his chest and in the other grasps handle of scimitar. 25 mm. (Pl. V-1)
HL. 7-212; Rm. 526; IM 63232.
Envelope; loan of silver; DF assigned to Sin-abushu (same as *Karim* DF 14).

10.
a. Enthroned "king" on fleece-covered stool on platform decorated with niches, wears cap with upturned brim. Two columns of faint inscription. Suppliant goddess behind bare-headed and beardless worshipper, wearing long garment with both hands clasped and held to chest. 24 mm. (Pl. II-5)
b. Suppliant goddess with elaborate multiple-horned crown, behind worshipper wearing long garment and cap with upturned brim. 22 mm.
c. Part of impression No. 9. Back part of throne with king and nude priest facing bull-eared god. 25 mm. (Pl. VII-5)
HL. 7-319; Rm. 526; IM 63299.
Envelope; loan contract; *Karim* 42; DF 12, of Sin-abushu date.

11.
a. Grooves made by seal caps above and below the scene. Suppliant goddess behind worshipper who offers kid to Sun God is ascending posture holding saw. Behind on platform is nude priest with pail and sprinkler. Two lines of inscription. 18 mm. (Pl. V-5)
b. Nude bearded figure with three pairs of curls holds clasped hands to chest, streams of water falling down from his waist to the ground. Ishtar in ascending posture, weapons sprouting from her shoulders, holds double lion-headed mace club in one hand and scimiltar in other. 22m. (Pl. VI-1)
HL. 7-87; IM 63134.
Envelope; DF same as No. 10.

12.
a. Nude priest with pail and sprinkler standing on platform, facing god wearing long garment, open down the front, holding scimitar in his left hand. In second group, worshipper with similar long garment, but wearing round cap raises one hand toward and presents kid to a deity not preserved in the impression. 20 mm. (Pl. VII-1)
b. In one group goddess Ishtar steps on lion; in second group, worshipper offers kid to Sun God holding saw and stepping on platform. 22 mm. (Pl. VI-2)
HL. 6-149; Rm. 447; IM 55386.
Envelope of Sin-abushu date.

13.
a. In one group, god in ascending posture carrying axe over his shoulder. In second group, suppliant

goddess behind worshipper who raises one hand while the other is wrapped inside his long garment which opens down the front. He wears slightly domed cap with pronounced brim. 20 mm. (Pl. XIV-3)

b. Nude priest with pail and sprinkler behind figure with mace facing suppliant goddess. 20 mm.
HL. 6-149; Rm. 447; IM 55376.
Envelope; loan contract, of Sin-abushu date.

14.
Nude hero with three pairs of curls seems to hold vase to his chest. Water flows from his shoulder into vase held by kneeling manikin. 17 mm. (Pl. VIII-8)
HL. 6-149; Rm. 447; IM 55374.
Envelope; loan contract of Sin-abushu date.

15.
a. Nude male figure wearing triplet girdle, and round cap, holds one hand to chest, extends other. 17 mm.
b. Suppliant goddess. 2 mm.
HL. 6-149; Rm. 447; IM 55380.
Envelope; Purchase of land; *Al-Hashimi* 35; of Sin-abushu date.

16.
a. Old Assyrian seal impression, showing worshipper behind suppliant goddess facing enthroned deity wearing round (Pl. XIV-1) "furry" cap or thick hair. In sky: star disc in crescent. 21 mm.
b. Back of nude figure, ball-staff. 15 mm.
c. Suppliant goddess behind worshipper offering kid to Sun God who steps on platform and holds saw. 22 mm.
d. Upper part of figure of suppliant goddess behind bareheaded worshipper. In sky: star disc in crescent. 11 mm.
e. Seated "king" wearing round cap with upturned brim and holding cup. Sun disc in crescent; part of column of inscription. 15 mm. (Pl. II-6)
f. Worshipper wearing long garment open down the front, and round cap with upturned brim, offering kid to enthroned god on platform who rests feet on reclining human-headed bull. Latter holds ring and staff. In sky: sun disc in crescent. 22 mm. (Pl. IV-2)
HL. 7-225; Rm. 526; IM 63240.
Envelope; sale of house; *Al-Hashimi* 12; of Sin-abushu date.

17.
a. Old Assyrian seal impression. Worshipper follows interceding goddess. 24 mm. (Pl. XIV-2)
b. Interceding goddess behind worshipper wearing long garment and round brimmed hat. One hand of worshipper is raised in supplication while with the other he offers kid to Sun God who steps on platform, holding saw (only his front is visible). 21 mm.
HL. 7-216; Rm. 526; IM 63236.
Envelope; adoption document; *Suleiman* 84; of Sin-abushu date.

18.
End of seal design showing back of figure (probably deity) facing left and small kneeling worshipper with brimmed hat, one hand raised to his mouth, facing right. 17 mm. (Pl. XIII-6)
HL. 7-217; Rm. 526; IM 63237.
Envelope; adoption document; of Sin-abushu date.

19.
Suppliant goddess, facing two crossed lion-griffins whose heads are turned backwards with open jaws in readiness to attack animal (lion) carried on its back over head of kneeling nude hero. To the left, lion crossed with a human-headed bull attacks a goat. In field: lizard above three drill holes, lowest attached to the body of an unidentified creature. Remains of lines for an inscription beside man-headed bull. 18 mm. (Pl. I-9)
HL. 7-92; L. III (heirloom); IM 63139.
Envelope; loan of silver; DF as No. 10.

20.
a. Conquering hero wearing short kilt and round cap, armed with curved weapon in one hand and brandishing seven-headed mace with other, trampling on kneeling victim. The hero is assisted by a second hero who seems to hold the victim by the head. Worshipper wearing long plain garment and round cap facing a god? Sun disc in crescent above. 18 mm. (Pl. VII-7)
b. Figure with mace stands behind nude priest with pail and sprinkler. In the field: bird above; below, lion-headed demon, brandishing weapon in one hand and holding head of human victim in other. 22 mm. (Pl. X-1)
c. God, holding scimitar, wearing long garment which opens down front, followed by nude attendant wearing slightly pointed cap and triple girdle, holding weapon. 16 mm. (Pl. VII-2)
d. Sun God? stepping on platform facing goddess Ishtar, who holds double lion-headed club and scimitar. 22 mm. (Pl. VI-3)
HL. 5-29; Rm 354; L.III (heirloom); IM 55353.
Envelope; purchase of house; *Al-Hashimi* 4; DF of Sin-abushu.

21.
a. Beardless god (or goddess) stands with one hand raised. 21 mm.
b. God with scimitar faces left, a crook before him and attendant holding weapon behind him. 16 mm.
HL. 5-38; Rm. 354; L. III (heirloom); IM 55153.
Envelope; business document; of Sin-abushu date.

22.
Suppliant goddess behind worshipper offering kid to Sun God, who is stepping on platform and holding saw horizontally. Behind him attendant holding object which is ending in knob, in other hand he may hold scimitar. 21 mm.
HL. 5-39; Rm. 354; L. III (heirloom); IM 55158.
Envelope; business document; of Sin-abushu date.

23.
Worshipper wearing long garment and cap, with one hand raised in supplication, led by interceding goddess toward enthroned god. Crescent in sky. Terminal: two columns of inscription; below, dog with collar. 19 mm. (Pl. III-8)
HL. 7-146; Rm. 501; IM 63218.
Envelope; letter.

24.
Presentation scene: worshipper, wearing long fringed garment and round cap, raising one hand in supplication, is led by interceding goddess toward enthroned figure on platform, holding cup. In field: vase above; below, ball-staff and mongoose. Terminal: two columns of inscription. 17 mm. (Pl. XXI-4)
HL. 6-149; Rm. 501; IM 63187.
Envelope; letter.

25.
a. Contest frieze, bull attacked by lion and bull-man, goat attacked by lion; nude hero with three pairs of curls wrestling with lion. 21 mm. (Pl. I-8)
b. Fragment of lion-griffin attacking human victim. 19 mm.
c. Nude figure facing left; god in short tunic holding mace, facing right, follows suppliant goddess. She stands behind worshipper wearing open garment and cap with upturned brim, who offers kid to Sun God in ascending posture holding saw. In upper field: part of fish creature? , three drill holes. Sun disc in crescent. 18 mm. (Pl. IX-2)
HL. 7-147; Rm. 501; IM 63185.
Envelope; legal document.

26.
Fragments of envelope.
a. End of an impression: goat standing upright on hind legs; below, seated lion? kneeling or seated figure. 22 mm. (Pl. XIII-7)
b. Bareheaded worshipper in long garment, facing enthroned "king" on platform decorated with niches, facing left wearing a cap with upturned brim and holding cup, is separated by two lines of inscription from suppliant goddess and bareheaded worshipper, both facing right. In sky: sun disc in crescent. 18 mm.
HL. 7-219; Rm. 531.
Envelope.

27.
a. Hero in combat position wearing short kilt and cap with upturned brim. Suppliant goddess behind worshipper with both hands clasped to chest facing enthroned figure on platform. In field: small figure of worshipper and monkey. 22 mm. (Pl. VII-6).
b. Suppliant goddess, behind worshipper facing enthroned Sun God, seated on large throne with high back. In field: vase, ball-staff, crescent and mongoose. Terminal: double lion-headed club-mace. 21 mm. (Pl. V-2)
c. Bearded nude hero with three pairs of curls wrestles with lion; kneeling human victim is attacked by lion-griffin; nude figure wearing triple girdle and round cap, holding up lion by hind leg, while placing foot on animal's head. In field: seated dog and human head. 22 mm. (Pl. I-4)
HL. 7-162; Rm. 531; IM 63197.
Envelope; sale of field.

28.
a. Small seal with grooves made by seal caps above and below scene. Worshipper faces enthroned deity. Crescent in sky; two columns of inscription. 10 mm.
b. Only remains of scene are suppliant goddess behind worshipper wearing cap. In field: vase above ball-staff. 20 mm.
HL. 7-245; IM 63247.
Fragments of envelope; letter.

29.
Fragment of scene; goddess leading worshipper of whom only hand is preserved, ball-staff in field. In sky: star in crescent. 22 mm.
HL. 7-342; IM 63295.
Envelope; loan contract.

30.
Presentation scene. Suppliant goddess behind

worshipper facing enthroned "king" seated on fleecy stool. Crescent in sky. Terminal: panel of inscription above dog. 21 mm. (Pl. II-12)
HL. 6-248; L. IV. IM 63295.
Envelope; purchase of field; *Suleiman* 40.

31.
a. Suppliant goddess behind worshipper facing enthroned deity (female goddess?), who has infant on lap. 16 mm. (Pl. III-6)
b. Part of an impression; enthroned figure, vase, staff of Ishtar's weapon, ball-staff, unidentifiable line. .08 mm.
HL. 6-85; Rm. 432.
Envelope; business document.

32.
a. Part of contest scene. Lion-griffin and parts of another animal. 20 mm.
b. Presentation scene. Skirt and feet of worshipper, facing enthroned "king," seated on fleece-covered stool; mongoose in field. 21 mm.
HL. 7-246; IM 63248.
Envelope; purchase of field. *Suleiman* 39.

33.
a. Side of head and arm of full-face goddess; bearded worshipper facing Sun God who holds saw in vertical position. Raised arm and outline and head of nude hero, holding lion over his head; second worshipper (perhaps from another impression). In field: crook and miniature worshipper. 16 mm. (Pl. V-3)
b. Fragmentary impression: head and shoulder of enthroned "king" holding cup. Sun disc in crescent; upper part of two columns of inscription. 07 mm. (Pl. II-7)
HL. 7-73; Rm. 499; IM 63125.
Envelope; loan contract.

34.
a. Worshipper behind figure with mace facing nude female. In field: vase, sun disc in crescent. 18 mm. (Pl. X-2)
b. God in ascending posture, facing Weather God who stands on bull, with large nude female standing beside him. Unidentified animal behind Weather God. 19 mm. (Pl. VIII-1)
HL. 6-2222; IM 55567.
Envelope; business document.

6.1.2. LEVEL III

35.
Lion-headed demon brandishing weapon with one hand and holding the head of an animal victim with other, facing god with scimitar and another weapon not preserved, trampling on human victim. Second group to right: nude priest with pail and sprinkler on double-tiered platform and standing behind worshipper pouring libation to god who is in ascending posture holding ring. Star disc in crescent. The slight repetition of the impression on the edge of figures is the result of the seal slipping. 22 mm. (Pl. IV-5)
HL. 7-133; Rm. 520; IM 63171.
Envelope; receipt; DF of Hammidashur.

36.
a. Part of an impression. Kneeling figure with one raised hand wearing round cap; goddess leading worshipper toward enthroned "deity" (completely missing). In sky: crescent, vase. 17 mm. (Pl. XXI-2)
b. Faint impression. Suppliant goddess with necklace counterweight behind bareheaded worshipper wearing long garment, facing enthroned figure on platform holding cup; mongoose behind throne; sun disc in sky. Two columns of inscription. 19 mm.
HL. 7-125; Rm. 520; IM 63163.
Envelope; loan contract; DF of Hammidashur.

37.
Impression of one seal; enthroned "king"? facing right, bull-man; nude priest with pail and sprinkler (missing). Vase above ball-staff. 15 mm. (Pl. II-11)
HL. 6-99; Rm. 87; IM 55398.
Tablet; business document; DF Ibalpiel I.

38.
The goddess Ishtar holding scimitar and double lion-headed club, with arrows protruding from quivers carried over her shoulders and her long hair ending down in curls. She places one foot on a lion. Suppliant goddess behind worshipper offering kid to Sun God holding saw and stepping on human-headed, seated bull. Sun disc in crescent. 24 mm. (Pl. V-8)
HL. 7-134; Rm. 520; IM 63172.
Envelope; sale of field; of Hammidashur date.

39.
a. Nude male figure with one raised hand, followed by second figure, only part of hand and leg are visible. 15 mm. (Pls. XVII-9, XXI-3)
b. Sun God with saw in ascending posture; conquering figure wearing cap, holding human head, facing god. 13 mm.
c. Suppliant goddess behind bareheaded worshipper facing enthroned "king" on platform. Behind the

throne is a kneeling nude figure with flowing vase. In sky: vase. 20 mm. (Pl. VIII-10)
HL. 7-144; Rm. 520; IM 63186.
Fragment of envelope.

40.
a. & b. Probably belonging to the same impression, showing worshipper with one raised hand and wearing long garment. a. has a vase in front of the worshipper; b. has a scorpion behind the second worshipper. 18 mm.
c. The legs of a human figure in a horizontal position. 05 mm.
d. Part of fighting scene. Nude hero seizing lion from behind. 15 mm.
HL. 7-133; Rm. 520; IM 63177.
Envelope; receipt; of Hammidashur date.

41.
a. Suppliant goddess with necklace counterweight, behind bareheaded worshipper faces enthroned "king" on platform, holding cup. Terminal: nude hero inverting lion. In field: monkey, drill holes and sun disc in crescent. 20 mm. (Pl. II-8)
b. Suppliant goddess behind worshipper wearing brimmed cap. 33 mm.
HL. 7-123; Rm. 520; IM 63161.
Envelope; loan of silver; of Hammidashur date.

42.
a. Figure with mace behind suppliant goddess who follows bareheaded worshipper toward enthroned "king" on platform who hold cup. Star disk in sky. 18 mm. (Pl. II-2)
b. Worshipper led by interceding goddess toward enthroned "king," wearing cap, seated on panelled throne. Terminal: crescent standard with tassels. In field: vase above ball-staff; crescent in sky. 20 mm. (Pl. II-3)
HL. 5-9; Rm. 341; IM 54686.
Envelope; purchase of field; of Hammidashur date; *Al-Hashimi* 25.

43.
Row of nude dancing men wearing round caps. Leader (appearing in the impression at the end of the row) has his forward hand raised, while the last dancer has one hand raised above his head. Small male figure as terminal. 23 mm. (Pl. XIV-8)
HL. 5-14; Rm. 341; IM 55155.
Envelope; purchase of field. *Al-Hashimi* 30; of Hammidashur date.

44.
Worshipper led by interceding goddess toward enthroned goddess. In sky: crescent. Inscription of two signs as terminal. 15 mm. (Pl. XXI-5)
HL. 5-19; Rm. 341; 54695.
Envelope; business document; *Suleiman* 46; of Hammidashur date.

45.
Lion-headed demon brandishing weapon and holding head of human victim, facing god with scimitar; dog balancing crook; worshipper with raised hand beside male figure (king), who embraces goddess. Cuneiform sign between them. 20 mm. (Pl. VII-3)
HL. 5-16; Rm. 341; IM 55162.
Envelope; sale of field; *Al-Hashimi* 26; of Hammidashur date.

46.
a. Enthroned "king" on platform, behind him attendant who seems to be holding an object over his shoulder. 20 mm.
b. Suppliant goddess, behind bareheaded worshipper, facing enthroned "king." In field: vase above ball-staff, crescent above mongoose. 18 mm.
c. God is ascending posture and holding a scimitar has right foot on lion. He faces left toward priest on double-tiered platform with pail and sprinkler (not visible), and toward goddess Ishtar, in profile. 20 mm. (Pl. VI-4)
HL. 5-15; Rm. 341; IM 54985.
Envelope; loan contract; *Karim* 41, DF No. 5. Contemporary with Hammidashur.

47.
a. Worshipper (king) enbracing female deity. Suppliant goddess. 21 mm. (Pl. XIII-4)
b. Contest scene; lion and lion-griffin attacking nude male; second male figure holding weapon, attacked by lion-headed demon. Snake as terminal. 15 mm. (Pl. XIV-7)
HL. 5-5; Rm. 341; IM 54684.
Envelope; business document.

48.
a. Hands of suppliant goddess behind priest with pail and sprinkler, both facing enthroned god. Sun disc in crescent above. Lion-griffin and lion attacking human victim. 19 mm. (Pl. XIV-5)
b. Suppliant goddess behind figure with mace. Crescent above. 21 mm. (Pl. X-3)
c. God enthroned on tortoise, water flowing from his shoulder; behind him kneeling nude attendant. Below figures runs stream of water with fish swimming in it. 17 mm. (Pl. VIII-6)
HL. 7-69; Rm 298; IM 54983.
Envelope; adoption document; DF of Naram-Sin.

49.
a. Stool for enthroned figure, only elbow visible. Lower part of pair of contestants, scorpion between. 12 mm. (Pl. I-12)
b. Suppliant goddess; inscription damaged. 14 mm.
HL. 7-179; Rm. 544; IM 63213.
Envelope; loan contract; *Karim* 34, DF No. 10 of Naram-Sin.

50.
a. God with mace (mace missing); star and fly above. 18 mm. (Pl. IX-3)
b. Terminal part of seal: scorpion above bow-legged dwarf, seated lion above mongoose, hedgehog below. 19 mm.
c. Upper part of figure with mace, facing suppliant goddess, with sun disc in crescent above. 23 mm. (Pl. X-4)
HL. 7-307; Rm. 525; IM 63252.
Envelope; contract; DF of Naram-Sin.

51.
a. Figure with mace, facing goddess who extends one hand, while folding other under garment. Fly in field above, ball-staff below. 16 mm. (Pl. IX-4)
b. Nude female. 18 mm. (Pl. XIII-3)
HL. 7-307; Rm. 525; IM 63254.
Envelope; DF of Naram-Sin.

52.
Worshipper offering kid. 23 mm.
HL. 7-307; Rm. 525; IM 63255.
Envelope; administrative document; DF of Naram-Sin.

53.
a. Goddess Ishtar, facing left, holding lion club and scimitar; worshipper offering kid to Sun God (only partly preserved) who, holding saw, steps on platform. 20 mm. (Pl. VI-8)
b. Figure with mace facing suppliant goddess. 20 mm. (Pl. X-5)
HL. 7 (register number is incorrect); IM 63256.
Envelope; loan contract; DF of Naram-Sin.

54.
a. Three lines of inscription, god with mace. 20 mm. (Pl. IX-5)
b. Figure with mace (mace missing), wearing a long curly beard, behind suppliant goddess. 20 mm. (Pl. IX-6)
c. Part of goddess Ishtar, two lines of inscription, back of bull-man. 22 mm. (Pl. VI-9)
d. Part of contest scene; hind leg of lion? Nude hero with three pairs of curls, overpowering bull, by grasping tail and placing one foot on animal's head. 22 mm.
e. Three lines of inscription; back of suppliant goddess. 26 mm.
f. Three lines of inscription; figure with mace. 22 mm. (Pl. X-8)
HL. 3-296; between L. III and II.
Envelope; adoption document; DF of Naram-Sin.

55.
a. Enthroned Water God holding flowing vase, front half only preserved. Water flows also from his shoulders and shoulders of kneeling nude attendant who has a triple-curl hairstyle. Several streams meet to flow into two round vases, and fish swim down the streams. The god rests foot on a fish-man, second fish-man is in field above, together with one large fish. 18 mm. (Pl. VIII-7)
b. Suppliant goddess behind worshipper offering kid. 20 mm.
HL. 7-307; Rm. 525; IM 63253.
Envelope; loan contract; of Naram-Sin date.

56.
a. Nude goddess with wings hanging from her waist. Figure of god with mace (mace missing). 19 mm. (Pl. IX-7)
b. Figure with mace facing suppliant goddess. Sun disc in crescent. 20 mm. (Pl. X-6)
HL. 7-307; Rm. 525; IM 63251.
Envelope; administrative document; of Naram-Sin date.

57.
a. Worshipper offering kid to enthroned god seated on stool with fleecy cover. 23 mm. (Pl. III-1)
b. Lion attacking human victim, suppliant goddess behind worshipper. In field: vase, and ball-staff below. 20 mm. (Pl. XIV-6)
c. God in ascending posture, nude female, hero with three pairs of curls in contest with animal, not preserved. Unidentified animal and DINGIR sign in field. 14 mm. (Pl. XIII-1)
HL. 7-154; Rm. 531; IM 63192.
Envelope; adoption document.

58.
Back only of enthroned deity on what appears to be tortoise, nude figure facing left holding stick or weapon; priest with pail and sprinkler, behind worshipper. 13 mm.
HL. 7-169; Rm. 531; IM 63203.
Envelope; administrative document.

59.
a. Grooves made by seal caps above and below scene. Suppliant goddess following worshipper

approaching enthroned god on platform, holding ring and staff. In sky: sun disc in crescent. Terminal: three lines of inscription. 15 mm. Pl. IV-3)

b. Enthroned king on platform; two lines of inscription; back of suppliant goddess. 19 mm.
HL. 7-357; Rm. 509; IM 63314.
Envelope; purchase of field; *Suleiman* 48.

60.

a. Enthroned god on fleece-covered stool, three lines of inscription, suppliant goddess wearing triple necklace. 30 mm. (Pl. III-2)
b. Bull-man holding standard (partly preserved, possibly ring post); worshipper led by interceding goddess toward enthroned "king" holding cup. Crescent in sky, seven drill holes scattered about in field. 15 mm. (Pl. III-3)
c. Lion-griffin facing left, worshipper led by interceding goddess toward enthroned deity only part of whose garment is visible. In field: ball-staff and miniature snake goddess. 21 mm. (Pl. III-4)
HL. 7-88; Rm. 505; IM 63135.
Envelope; sale of field; *Al-Hashimi* 14.

61.

God stepping on platform, holding ring. Bull-man in contest with nude hero with three pairs of curls, lion-fish between them. 22 mm. (Pls. I-11, IV-6)
HL. 2-560; Rm. 142; IM 51609.
Jar sealing.

62.

a. Figure with mace facing suppliant goddess; between them, sun disc in crescent above small figure of worshipper. 19 mm. (Pl. X-7)
b. Nude female, god with mace? to right. 16 mm. (Pl. IX-8)
c. Suppliant goddess, following worshipper. 18 mm.
d. The goddess Ishtar carrying her weapons. Suppliant goddess behind. In field: part of lion-fish?, sun disc in crescent. 22 mm. (Pl. VI-5)
HL. 5-25; Rm. 320; IM 54991.
Tablet; administrative document.

63.

a. Legs of male figure, suppliant goddess behind worshipper of whom only part is showing. In field: vase above ball-staff. 20 mm.
b. Suppliant goddess behind worshipper facing enthroned "king" holding cup. In sky: sun disc in crescent. 11 mm. (Pl. II-9)
HL. 7-350; Rm. 544.
Fragment of an envelope.

64.

God with mace? facing suppliant goddess; lion-fish between them; to right, goddess standing full-face. (Pl. IX-9)
HL. 7-177; Rm. 544; IM 63268.
Envelope.

65.

a. Nude priest with pail and sprinkler behind worshipper facing Sun God in ascending posture holding saw. 21 mm.
b. Sun God stepping on platform holding saw; nude female behind him; figure of nude male with round brimmed cap. In sky: star, inscription. 19 mm. (Pl. XIII-2)
HL. 7-180; Rm. 544; IM 63214.
Envelope; adoption document.

66.

Two impressions belonging to one seal. Combat scene; goat attacked by lion. Bull-man wrestling with nude hero, human victim overpowered by lion griffin. Fish between legs of bull-man and hero. 20 mm. (Pl. I-5)
HL. 5-31; Rm. 353.
Lump of clay.

67.

a. Worshipper facing Sun God in ascending posture with saw. 19 mm.
b. Figure with mace? Sun disc in crescent. 13 mm.
c. Fish. 03 mm.
HL. 3-140; L. III; IM 52688.
Envelope; adoption document.

68.

Striding male figure, and two crossed dancers? 15 mm.
HL. 9-10; IM 67009.
Envelope.

69.

a. Front of suppliant goddess, bareheaded worshipper, his clasped hands held to his chest, facing enthroned "king" on platform. Terminal: two columns of inscription. 22 mm. (Pl. II-10)
b. Enthroned god holding ring and staff; sun disc in crescent; one faint column of inscription. 20 mm. (Pl. IV-7)
c. Worshipper led by interceding goddess toward an enthroned god. Terminal: two columns of inscription. 17 mm. (Pl. III-9)
d. Front of suppliant goddess, behind worshipper with thick furry hair (hat?); between them star, vase below. 17 mm.
e. Lower front of deity, front of monster. 10 mm.
f. Sun god, holding saw and stepping on artificial mountain; to right, nude priest holding curved object and standing on double-tiered platform;

cuneiform signs scattered in the field. 18 mm. (Pl. IV-4)

g. Front of worshipper offering kid to Sun God who is in ascending posture holding saw (only front visible). 11 mm.
h. Lower part of enthroned "king," and figure with mace. 10 mm.
i. Goat grasped by hero with thick furry hair (hat). 15 mm.
j. Nude hero with three pairs of curls, goat with head turned back, attacked by a second hero? 10 mm.
k. Part of victim grasped by lion-griffin. 12 mm.
l. Bull held by human-headed bull. 18 mm.
m. Two lions standing on their hind legs facing each other. Lion-griffin holding moon standard. 07 mm. (Pl. I-13)
n. Small male figure between two other figures. 06 mm. (Pl. XIII-5)

HL. 7-144; Rm. 520.
Fragments of envelopes.

6.1.3. LEVEL II

70.
Worshipper with round cap and long garment that opens down the front. Two columns of inscription. 20 mm.
HL. 2-484; Rm. 155; IM 51331.
Tablet; DF of King Dadusha; *Baqir* No. 9.

71.
Back of god, behind him is humped bull on its hind legs. Lion-fish above. 22 mm.
HL. 4-203; Rm. 252; IM 54446.
Tablet; administrative document; DF of Ibalpiel II, 1st year; *Baqir* No. 21.

72.
Two columns of inscription, suppliant goddess. 23 mm.
Rm. 190; between L. III and II; IM 52601.
Tablet; Ibalpiel II, 5th year; *Baqir* 25.

73.
Figure with mace; three columns of inscription. 20 mm. (Pl. X-9)
HL. 2-77; Rm. 55; IM 51177.
Tablet; DF of Ibalpiel II, 7th year; *Baqir* 27.

74.
Part of the figure with mace; two columns of inscription. 20 mm. (Pl. X-10)
HL. 2-244; Rm. 133; IM 51195.
Tablet; loan of sesame; DF of Ibalpiel II, 6th year; *Baqir* 28.

75.
Three columns of inscription, back of figure with mace. 30 mm. (Pl. X-11)
Tablet; DF of Ibalpiel II, 8th year; *Baqir* 28a.

76.
a. Worshipper (king), wearing a high cap and open fringed mantle, led by a goddess to an enthroned deity holding ring and staff and resting its feet on a dragon whose body extends underneath the throne. Above are eight-pointed star and crescent. Terminal: three columns of inscription. Rolled all over envelope. 20 mm. (Pl. IV-14).
b. Worshipper offering kid to Sun God stepping on platform and holding his saw. Behind him is one leg of nude attendant with flowing vase; above: star disc in crescent. Rolled on tablet. 20 mm.
c. God (Sun God), followed by suppliant goddess; above, star disc in crescent, vase. Rolled on tablet. 15 mm.

HL. 3-129; Rm. 180; IM 52599.
Tablet and envelope; division of an estate; DF of Ibalpiel II, 10th year; *Baqir* 30; Ellis, *JCS* 26, 1974.

77.
a. Suppliant goddess (left) back-to-back with figure with mace. In the proper rolling the figures should be facing. 26 mm. (Pl. XI-1)
b. Kneeling nude hero with curls, in profile. 21 mm.

HL. 3-103; Rm. 180; IM 52626.
Tablet; legal document; DF of Ibalpiel II, 10th year; *Baqir* 30a.

78.
Figure with mace facing suppliant goddess. Between them: upper field, disc in crescent; below, crook. 22 mm. (Pl. XI-2)
HL. 2-298; Rm. 143; IM 51741.
Tablet; Ibalpiel II, 11th year; *Baqir* 31.

79.
Figure with mace facing suppliant goddess. Two columns of inscription. The original seal had a metal cap with filigree decoration. 27 mm. (Pl. XI-5)
HL. 2-257; Rm. 136; IM 52178.
Envelope; letter of Ibalpiel II date.

80.
a. Figure with mace (missing), porcupine in upper field, fish below. 20 mm. (Pl. XI-3)
b. Three columns of inscription. 26 mm.

HL. 2-318; Rm. 133.
Tablet; administrative document; DF of Ibalpiel II; *Baqir* 34.

81.
Only three columns of inscription are visible. 28 mm.
Courtyard 110; IM 52275.
Tablet; DF of Ibalpiel II; *Baqir* 15.

82.
Worshipper with slightly domed brimmed cap. Three columns of inscription. 20 mm.
HL. 2-295; Rm. 143; IM 51788.
Tablet; DF of Ibalpiel II; *Baqir* 33.

83.
Seal of Ibalpiel II, with metal filigree caps, showing only four columns of inscription. 21 mm. (Pl. XI-4)
HL. 2-379; Rm. 133; IM 51251.
Envelope; letter.

84.
Three columns of inscription. Back of figure with mace. 20 mm.
IM. 51281.
Tablet.

85.
Worshipper facing god, holding ring and staff, who places his foot on a dragon. Behind the god is a nude attendant with three pairs of curls. Terminal: three columns of inscription. 20 mm. (Pl. IV-8)
HL. 2-435; Rm. 133; IM 51289.
Envelope; Ibalpiel II date.

86.
a. Probably the same impression as no. 85. 18 mm.
b. Part of prancing animal; head of lion. 11 mm.
c. Contest scene; animal, lion attacking kneeling human victim, another lion beyond. The seal was probably unfinished as apparent from the prominence of drill holes in many parts. 15 mm.
HL. 2-501; Rm. 133; IM 51658.
Envelope; Ibalpiel II date.

87.
a. Upper part of two columns of inscriptions. 10 mm.
b. Figure with mace. 15 mm.
c. Nude hero with three pairs of curls, in contest posture. 25 mm.
HL. 2-244; Rm. 133; IM 51190.
Tablet; legal document; Ibalpiel II date.

88.
a. Figure with mace, facing suppliant goddess; two columns of inscription. 25 mm. (Pl. XI-6)
b. Only a fragment of the filigree cap is visible. (Pl. XI-10)
HL. 2-241; Rm. 133; IM 51265.
Envelope; letter; Ibalpiel II date.

89.
Three columns of inscription. Figure with mace, facing suppliant goddess. The seal impression had grooved edges. 25 mm. (Pl. XI-7)
HL. 2-318; Rm. 133; IM 51657.
Envelope; Ibalpiel II date.

90.
Three columns of faint inscription. Suppliant goddess faces Weather God who holds lightning fork and stands on humped bull. In field: tortoise. 25 mm. (Pl. VIII-3)
HL. 2-318; Rm. 133.
Envelope; letter; Ibalpiel II date.

91.
Figure with mace facing suppliant goddess; star in sky. 17 mm. (Pl. XI-4)
HL. 2-244; Rm. 133; IM 51179.
Tablet; contract; Ibalpiel II date.

92.
Same impression as No. 82.
IM 51600.
Envelope.

93.
a. Capped seal; three columns of inscription; suppliant goddess. 25 mm.
b. Only the impression of the filigree metal cap remains. 25 mm. (Pl. XI-11)
Rm. 143; IM 51999.
Envelope; business document; Ibalpiel II date.

94.
Three columns of faint inscription. Figure with mace (head missing), facing suppliant goddess. 24 mm. (Pl. XI-8)
HL. 2-318; Rm. 143; IM 51542.
Envelope; receipt; Ibalpiel II date.

95.
Figure with mace. Two columns of inscription. 25 mm. (Pl. XII-5)
HL. 2-303; Rm. 143; IM 51203.
Tablet; administrative document dealing with beer; Ibalpiel II date.

96.
Suppliant goddess. Three columns of inscription; part of figure with mace. 18 mm.
HL. 2-246; Rm. 143; IM 52188.
Tablet; record of beer; Ibalpiel II date.

97.
A god? In the field: two fish and hedghog. 22 mm.
HL. 1-1; Rm. 5; IM 51057.
Tablet; loan contract.

98.
Two columns of faint inscription. Figure with mace. 25 mm. (Pl. XII-6)
HL. 9-52; Rm. 620; IM 67340.
Envelope.

99.
a. Three columns of inscription. Figure with mace facing suppliant goddess. In sky: sun disc in crescent. 20 mm. (Pl. XII-4)
b. Suppliant goddess, facing Sun God in ascending posture. In sky: sun disc. 20 mm.
c. Probably the goddess Ishtar holding scimitar. 20 mm.
HL. 2-31; Rm. 77; IM 51349.
Envelope; adoption document.

100.
Only the lower section of a seal. Figure with mace, facing suppliant goddess. Between them, small figure of worshipper. Worshipper behind a second goddess. 13 mm. (Pl. XIII-8)
HL. 2-33; Square 9; IM 53913.
Envelope; letter.

101.
Figure (nude?) with one hand raised, followed by a second nude figure rendered with stick-like limbs raising left hand over its head. Worshipper offering kid to Sun God holding saw, only lower half of figures remains. 15 mm. (Pl. XIV-10)
HL. 4-333; Rm. 265; IM 54340.
Envelope; administrative document.

102.
a. God in ascending posture. In sky: sun disc in crescent. 18 mm.
b. Small male figure raising one hand placed in upper field; beside it are two columns of inscription. 15 mm.
c. God in ascending posture to left. Figure with mace partly shown, turned to right; between them cuneiform inscription. 21 mm.
HL. 4-268; Rm. 252; IM 52986.
Tablet; legal document.

103.
Figure with mace between two suppliant goddesses. 24 mm. (Pl. XII-7)
HL. 4-431; Rm. 297; IM 54029.
Envelope; letter.

104.
a. Goddess, full-face, standing on goat-fish (only left side visible); facing her is suppliant goddess. Disc in crescent above fish. 20 mm. (Pl. VIII-9)
b. God in ascending posture. 06 mm.
c. Mongoose. 04 mm.
HL. 6-48; Rm. 329; IM 55433.
Envelope; adoption document; *Suleiman* 82.

105.
Three columns of inscription. Back of suppliant goddess. 25 mm.
HL. 3-17; Square 11; IM 51426.
Tablet; receipt.

106.
Worshipper raising one hand in supplication, possibly facing enthroned "king" (only hand holding cup is visible). Several motifs in field: small figure of worshipper above nude female, second small figure of worshipper above lion club, porcupine below. 23 mm. (Pl. XIII-9)
HL. 4-9; IM 52889.
Envelope; administrative document.

107.
Three columns of inscription; figure with mace. Seal had metal caps. 28 mm. (Pl. XII-8)
HL. 2; IM 51546.
Envelope.

108.
a. Weather God on dragon. Figure with mace. 22 mm. (Pl. VIII-5)
b. Leg of human figure. Goddess Ishtar stepping on lion. 10 mm. (Pl. VI-6)
c. Two columns of inscription. Suppliant goddess. 23 mm.
HL. 3-2; Rm. 171; IM 52268.
Envelope; loan contract; DF of Naram-Sin.

109.
a. Enthroned deity. Disc in crescent sky. 24 mm. (Pl. XIV-9)
b. Worshipper before enthroned deity. Left of worshipper two striding men separated by dividing line from lion below. 14 mm. (Pl. XIV-9)
HL. 4-65; Rm. 240; IM 52950.
Envelope; contract.

110.
Three columns of inscription, back of figure with mace. 20 mm.
HL. 2; IM 51281.
Envelope.

111.
God holding crook (Amurru), follows second god holding mace. Both wear high square hats and face Weather God in ascending posture holding lightning fork. Crook in field. 20 mm. (Pl. VIII-4)
HL. 2-408; Rm. 101; IM 52138.
Bulla.

112.
Nude male figure behind worshipper led by

interceding goddess to enthroned deity. In field: vase above ball-staff, crescent. 08 mm.
HL. 4-44; Rm. 235; between L. III and L. II; IM 52944.
Strip of clay; tag?

113.
Part of figure with mace (mace missing), facing suppliant goddess (head missing), star between them. 18 mm. (Pl. XII-3)
HL. 2-395; Rm. 122; IM 51230.
Bulla.

114.
a. Figure with mace, facing suppliant goddess. Star in sky. 23 mm. (Pl. XII-2)
b. Incomplete contest scene: nude hero, second nude hero lifting lion over his head; lion standing on hind legs (same impression as 115). 25 mm.
c. Worshipper, behind figure with mace facing suppliant goddess. Terminal: lion attacking goat. In the field: star, vase above lightning fork, snake, and ball-staff, miniature enthroned god holding curved weapon (saw?), fish, and small figure of worshipper. 24 mm. (Pl. XII-1)
HL. 2-33; Square 9; IM 51622.
Envelope.

115.
Three pairs of contestants. Nude hero with curls attacking lion from the back; lion in inverted position held by second hero; third hero in profile, raising a lion in the air. 25 mm. (Pl. I-6)
HL. 3-21; Square 110.
Fragment of envelope.

6.2. Tell Dhiba'i

116.
God in ascending posture. Three columns of inscription of irregular width. Seal with caps. 13 mm. (Pl. XIX-1)
Envelope; Dh. 1-1; Temple L. V; IM 52787.
Loan contract; DF contemporary to Hammi-dashur; *Baqir* Ch. 3; *Karim* no. 6 has similar impression.

117.
Male figure raising one hand, behind priest with pail and sprinkler; both wear kilts with belts. Two columns of inscription. 20 mm. (Pl. XVII-7)
Temple; L. V; IM 52782.
Envelope; loan contract.

118.
God with mace, attendant holding a curved weapon behind figure with mace, facing god in ascending posture. 17 mm. (Pl. XIX-2)
Temple L. V; IM 52772.
Envelope; loan contract.

119.
Male figure wearing cap marked with parallel lines, triple girdle and short kilt, and raising one hand in supplication stands behind worshipper wearing a similar cap, long garment and raising right hand in front of his face, other hand is hidden by garment. One column of inscription. Slipped impression. 18 mm. (Pl. XVII-6)
Rm. 106; L. V; IM 66978.
Fragment of envelope.

120.
Elaborate contest frieze. Two nude heroes with triple curls have a lion between them. Second group: lion-griffin attacking goat who rests its foreleg on shoulder of a hero overpowering a bull by holding it upside down. 21 mm. (Pl. XX-5)
Temple; L. V; IM 52770.
Envelope; loan contract.

121.
Two columns of inscription. Nude male figure, wearing round cap with upturned brim. 20 mm. (Pl. XVII-10)
L. V.
Fragment of envelope.

122.
Worshipper with one hand raised in front of his face, facing enthroned "king" on panelled stool, holding cup? Tree in stand behind worshipper. 20 mm. (Pl. XVII-12)
L. V; IM 52853.
Fragment of envelope.

123.
a. Nude male figure wearing round cap and raising one hand, behind worshipper, also raising one hand in supplication. He, in turn, is behind a third worshipper who is bareheaded and has both hands clasped to his chest. All face enthroned deity? In field: crescent above mongoose. 12 mm. (Pl. XV-1)
b. Suppliant goddess faces the goddess Ishtar who

steps on lion and holds double lion club and scimitar. 20 mm. (Pl. XVIII-8)
Rm. 29 mm; L. I; IM 52872.
Lump of clay; bulla?

124.
a. Four drill holes; back of suppliant goddess. 15 mm.
b. Lion-griffin turned 90% menacing kneeling nude male, crescent and scorpion in the field. 15 mm. (Pl. XX-12)
Rm. 160; L. V; west of Temple; Dh. 2-442; IM 67273.
Envelope; receipt.

125.
Worshipper led by second worshipper toward enthroned "king" holding cup. 8 mm. (Pl. XVI-2)
Dh. 1-3; Rm. 120; L. V; IM 66981.
Envelope.

126.
Worshipper approaching enthroned "king." Behind king is a large figure of the bow-legged dwarf, and a lion club. In field: crescent above vase. 20 mm. (Pl. XVI-1)
Dh. 2-574; Rm. 120; L. V.
Fragment of envelope.

127.
Two worshippers face enthroned "king." Crescent in sky. 15 mm.
Dh. 2-571; Rm. 120; L. V.
Fragment of envelope.

128.
a. Worshipper facing enthroned deity holding cup. Vase above ball-staff. 15 mm.
b.1. Worshipper with one raised hand behind second worshipper facing enthroned deity on platform holding cup. In field: sun disc in crescent above ball-staff. Two columns of inscription. 14 mm.
b.2. Same impression with seal recut as worshipper acquired long hair.
c. Combat scene; lion attacking human victim; two more lions, between them a goat. 20 mm. (Pl. XX-9)
d. Two pairs of contestants: bull-man restraining lion; nude hero with curls holding goat. Terminal: lion club. 21 mm. (Pl. XX-1)
e. Lion-griffin attacking human victim. 12 mm. (Pl. XX-10)
f. Two crossed human-faced bulls, one is restrained by a nude hero. 15 mm.
g. Worshipper led by interceding goddess toward enthroned god holding cup. Single cuneiform sign in one column. In field: unidentifiable object, vase above ball-staff. 19 mm. (Pl. XV-8)
h. Worshipper led by interceding goddess toward enthroned deity holding cup. In field: crescent above bird; vase. 14 mm. (Pl. XV-9)
i. Suppliant goddess facing enthroned "king." Crescent in sky. Part of star, and faint columns of inscription. 8 mm.
j. Bareheaded worshipper facing enthroned "king," holding cup. Star disc in crescent in sky. 21 mm. (Pl. XVIII-1)
k. Worshipper (head missing), behind second worshipper. In the field; crescent above ball-staff. 15 mm.
l. Partial outline of a bull man. Suppliant goddess behind female worshipper facing enthroned "king," holding cup. Two panels for inscription; bull-man. In field: star disc above spread-winged eagle. 24 mm. (Pl. XVIII-3)
m. Four columns of inscription; suppliant goddess behind female worshipper who wears necklace with counterweight. 21 mm. (Pl. XVIII-4)
n. Suppliant goddess, behind worshipper facing enthroned god holding cup. Star disc in crescent in sky. Two columns of inscription as terminal. 20 mm.
o. Worshipper led by interceding goddess toward enthroned "king" on (fleeced) stool. In sky: disc in crescent. 16 mm. (Pl. XV-10)
p. Interceding goddess leading worshipper (only hand of latter remains visible), toward enthroned "king" holding cup. Between them ball-staff. 11 mm. (Pl. XV-11)
Dh. 2-440; L. V; Rm. 120.
Fragments of envelopes.

129.
a. Presentation to enthroned "king" holding cup, toward whom suppliant goddess follows bareheaded worshipper. Terminal: short-haired bearded hero overpowering bull, three drill holes in field, mongoose below. 21 mm. (Pl. XVIII-2)
b. Suppliant goddess behind worshipper, three drill holes between them in sky. 23 mm.
c. Back of enthroned deity on platform. Behind throne is an attendant carrying towel(?) hanging over forearm and holding a curved object (fly whisk) with the other. Two lines of inscription. In field: crouching monkey. 19 mm. (Pl. XVII-4)
d. Presentation to enthroned "king," nude bearded figure with three pairs of curls holding vase? with water flowing down from his waist. Suppliant goddess behind traces of worshipper belonging to presentation scene. 15 mm. (Pl. XVIII-6)
e. Nude figure, upper part missing. 15 mm. (Pl. XVII-8)

f. Enthroned god on a platform holding ring and staff, receiving the offering of a kid, probably from a worshipper. Four columns of inscription. 17 mm. (Pl. XVIII-7)
g. Two scenes of ritual marriage, one standing and one horizontal. In field: scorpion, star, and the horns of a goat are preserved. 15 mm. (Pl. XIX-3)
h. See description for 123.a (Pl. XV-2)
i. Bareheaded worshipper, led by interceding goddess toward enthroned "king" on platform, holding cup. Crescent in sky. Lion club as terminal. 11 mm. (Pl. XV-13)
j. Dog? in front of enthroned god (only feet preserved) on platform. 13 mm.
k. Combat scene. Lion attacking nude hero with three pairs of curls who is grasping a bull. Bull-man grasping lion. 10 mm. (Pl. XX-6)
l. Lion-headed winged demon. 18 mm.
m. Fragment. Upper part of a suppliant goddess wearing necklace with counterweight; behind worshipper wearing cap with brim.
n. Part of head and shoulder of worshipper, facing enthroned "king" holding cup. Crescent in sky. Vase in field. 15 mm. (Pl. XIX-7)
o. Faint impression. Enthroned "king" on fleeced stool, resting his feet on platform. One column of inscription. Suppliant goddess. 15 mm.
p. Indistinct object. Part of arm and leg of bow-legged manikin. 14 mm.
L. V; Rm. 29; IM 54045.
Fragments of envelopes.

130
a. Probably a scorpion-man on a platform, back of worshippers. 11 mm. (Pl. XIX-5)
b. Nude male figure carrying a bow behind worshipper facing spear or spade; second worshipper. Mongoose in field. 17 mm. (Pl. XVII-5)
c. Human victim, kneeling on double-tiered platform, attacked by two lions. 15 mm. (Pl. XX-11)
d. Suppliant goddess, possibly standing behind the deity. Three columns of inscription; bareheaded worshipper, his hands folded to his waist, facing enthroned deity on platform. 20 mm. (Pl. XIX-6)
e. Bull-man fighting with lion; nude hero inverting second lion. In field: seated dog, monkey to left. On right, walking bird, seated goat, and eight drill holes. 17 mm. (Pl. XX-8)
f. Parts of two bull-men on either side of small figures of two lions clutching at standard (one head missing). 15 mm. (Pl. XV-6)
g. Combat scene. Lion; second lion being attacked by nude hero and bull-man. Lion-griffin attacking human victim. (Impression c above may belong to same seal.) 15 mm. (Pl. XX-7)
h. Enthroned deity; one column of inscription; worshipper. 20 mm.
i. Worshipper behind second worshipper, led by interceding goddess facing enthroned "king" holding cup. Crescent in sky. 17 mm. (Pl. XV-12)
j. Enthroned deity. Double lion-headed club-mace with tassels. Worshipper led by interceding goddess. In field: crescent star and ball-staff. 13 mm. (Pl. XV-15)
k. Two figures, between them palm tree in stand; bow-legged dwarf. 13 mm. (Pl. XVII-1)
l. Two male worshippers. Star in field. 19 mm.
m. Two lions on either side of standard. Worshipper facing enthroned "king" holding cup. Crescent above star. 10 mm. (Pl. XV-5)
Dh. 2-441; Rm. 120; L. V.
Fragments of envelopes.

131.
a. Worshipper behind second worshipper facing enthroned "king" on platform holding cup. Behind king are two nude figures wrestling? In field: star above ball-staff; crescent above mongoose; vase. Terminal: snake. 20 mm. (Pl. XXI-7)
b. Old Assyrian seal; worshipper behind god. 15 mm. (Pl. XIX-4)
c. Suppliant goddess behind worshipper facing enthroned "king" holding cup. In field: crook, disc in crescent above snake. Terminal: three columns of inscription. The seal had grooved edges. 17 mm. (Pl. XVIII-5)
d. Suppliant goddess behind worshipper facing enthroned "king" holding cup. In field: lizard; disc in crescent above vase, below ball-staff. Terminal: two columns of inscription. 15 mm. (Pl. XVI-4)
e. Worshipper facing enthroned "king" on platform; bow-legged dwarf in between. Crescent above. 22 mm.
f. Row of three suppliant goddesses facing enthroned "king" holding cup. In field: lizard, vase above ball-staff, sun disc in crescent above star. Terminal: lizard being eaten by snake. 20 mm. (Pl. XVI-6)
g. Suppliant goddess behind worshipper facing enthroned "king" holding cup. In field: mongoose; vase above ball-staff. Terminal: two columns of inscription. 17 mm. (Pl. XVI-5)
h. Suppliant goddess behind worshipper facing enthroned "king" holding cup. In field: star above bird; crescent above ball-staff. Terminal: plant. 16 mm. (Pl. XVI-3)
i. Bow-legged dwarf. Two male figurines wearing caps on either side of bird? supporting a crescent standard. 13 mm. (Pl. XVII-3)
j. Two male figures on either side of tree in vase; nude figure to right. 12 mm. (Pl. XVII-2)
k. Enthroned "king." Column of inscription.

Bull-man in contest with nude hero? to right, probably worshipper whose skirt is preserved. 24 mm. (Pl. XX-2)

l. Enthroned god in presentation scene, two columns of inscription, and suppliant goddess. 30 mm.

m. Human-headed bull held by bull-man. Remains of two columns of inscription. 25 mm. (Pl. XX-3)

n. Male figure in short kilt (in contest action with lion). Worshipper facing enthroned "king" holding cup. Behind him, lower part of male figure who seems to be holding down a scimitar. 15 mm. (Pl. XV-7)

o. Column of inscription, wrestling bull-man; ball-staff. 22 mm. (Pl. XX-4)

p. Enthroned "king"; star in sky. 23 mm.

q. Enthroned "king" holding cup. Crescent in sky. Ball-staff and vase in field, elbow of worshipper. 15 mm.

r. Worshipper led by interceding goddess toward enthroned "king" holding cup. In field: star above scorpion; crescent above bird. 22 mm. (Pl. XV-14)

s. Worshipper led by interceding goddess (head missing) toward enthroned "king." Terminal: double lion-headed mace (incorrectly cut) or club with tassels. In field: back of fish above ball-staff; crescent above vase. 12 mm. (Pl. XV-16)

t. Enthroned "king" on panelled stool, giant lizard as terminal. 16 mm. (Pl. XVII-11)
Dh. 2-450; Rm. 120; L. V.
Fragments from envelopes.

6.3 The Archive of Nurshamash

132.
Impression incomplete, due to seal being rolled at intervals. Contest frieze. Back of nude hero with three pairs of curls, back of lion. Lion contesting with bull-man. Star between them. Winged lion dragon. Crescent in sky. 21 mm.
IM 43642, (*Raschid* 54); IM 43643 (68).
Envelope.

133.
Fragment of impression. Nude male figure with round cap in striding posture, raising one hand. Part of second figure visible, possible worshipper. 20 mm.
IM 43555 (*Raschid* 54); IM 43819 (74); IM 43866 (43); 43883 (77).
Envelope.

134.

a. Incomplete impression. Worshipper in long garment pouring libation in front of Sun God who places one foot on platform and holds saw. Star disc in crescent above three drill holes. 26 mm. (Pl. V-9)

b. Possibly continuation of 134.a. Nude priest with pail and sprinkler behind worshipper. Terminal: male figure wearing round cap and short kilt, engraved in reversed position. 26 mm.
IM. 43556 (*Raschid* 102); IM 43629 (16); IM 43671 (97); IM 43717 (99).
Envelope.

135.

a. Priest wearing short garment extending sprinkler to male figure wearing round brimmed cap and short garment, having both hands clasped in front of his chest. Two columns of faint inscription. 19 mm. (Pl. XXI-6)

b. God holding scimitar. To the right are vase and ball-staff. Behind and to his left is conquering king, who holds multiple mace, brandishing curved weapon, and steps on leg of fallen victim. Head of second heroic figure preserved. 26 mm. (Pl. VII-8)

c. Enthroned "king" on platform. Human head in field. 19 mm.

d. God in ascending posture placing one foot on platform. Terminal: two columns of inscription. 23 mm. (Pl. XXI-7)
IM 43559 (*Raschid* 98).
Envelope.

136.
Part of an impression. Worshipper wearing long garment, and round cap with brim, carrying a sacrificial animal. 22 mm.
IM 43560 (*Raschid* 67).
Envelope.

137.

a. Worshipper wearing round cap and long garment, raises one hand while other is hidden by garment. He faces god who holds scimitar in lowered left hand. Eight-pointed star in sky. 14 mm. (Pl. VII-4)

b. Probably same impression as 137.c. Here the quivers behind the goddess' shoulders are visible as are the pleats of her long skirt. 25 mm. (Pl. VI-10)

c. Figure with mace facing goddess Ishtar, who holds

double lion-headed club and scimitar and steps on her lion; priest with pail and sprinkler stands on platform. 25 mm.
IM 43562 (*Raschid* 10); IM 43849 (23).
Envelope.

138.
Suppliant deity with elaborate crown behind worshipper, who wears bracelet on his raised hand and faces Sun God holding saw. Terminal: nude hero fighting with lion. Old Assyrian impression. 23 mm. (Pl. XIX-4)
IM 43591 (*Raschid* 57).
Envelope.

139.
War Goddess with head in profile, doubtless receiving, in presentation scene, worshipper with one raised hand, wearing round cap and long garment opening down the front. In sky: star disc in crescent. 20 mm. (Pl. VI-7)
IM 43654 (*Raschid* 34).
Envelope.

140.
a. Outline of back possibly of goddess Ishtar holding scimitar, followed by nude priest with pail and sprinkler. 25 mm.
b. Sun God in ascending posture. Behind him goddess full-face wearing single horned crown and flounced garment. Between them inscription of DINGIR Shamash. Part of a suppliant goddess visible to right. 20 mm. (Pl. XXI-8)
IM 43665 (*Raschid* 38); IM 43843 (22); IM 43900 (62).
Envelope.

141.
a. Worshipper offering kid, long garment open in front. Crescent above, crook behind him and disc in crescent in upper field before him. 24 mm.
b. Part of full-faced goddess, wearing single horned crown, and long founced garment. Two bow-legged manikins, one above the other, the upper in reversed position. 25 mm.
IM 43666 (*Raschid* 21).
Envelope.

142.
a. Presentation scene; suppliant goddess, and worshipper offering kid face Sun God in ascending posture holding saw. 22 mm.
b. Lower half of the goddess Ishtar, holding club and rein of lion on which she steps. 16 mm.
IM 43668 (*Raschid* 13).
Envelope.

143.
Figure with mace facing suppliant goddess. 18 mm.
IM 43684 (*Raschid* 93).
Envelope.

144.
God holding scimitar. Box of inscription, below scorpion; back of worshipper is visible. 22 mm. (Pl. XXI-9)
IM 43688 (*Raschid* 94); IM 43711 (39); IM 43902 (26).
Envelope.

145.
Front of suppliant goddess behind worshipper who stands with hands clasped at waist. He wears a long garment with one shoulder bare. Enthroned "king" with round cap on panelled stool holds cup. In field: vase above ball-staff, crescent above two drill holes above bow-legged manikin. 20 mm.
IM 43818 (*Raschid* 27); IM 43843 (22).
Envelope.

146.
a. Legs only of attendant (priest) standing on platform decorated with scale pattern. 10 mm.
b. Male nude figure (head missing). Inscription in field. 20 mm. (Pl. XXI-11)
IM 43819 (*Raschid* 74).
Envelope.

147.
God wearing long garment that covers one bent arm and has opening in front. In his lowered hand he holds scimitar. Conquering king in short kilt holding multiple mace and trampling over enemy. Worshipper. 17 mm. (Pl. VII-9)
IM 43827 (*Raschid* 75).
Envelope.

148.
Faint impression. God behind female deity. In sky: human head. 22 mm.
IM 43843 (*Raschid* 22); IM 43818 (27); IM 43665 (38).
Envelope.

149.
Bearded "god" wearing flounced garment offering kid to Sun God, in ascending posture. Between them, above, cow suckling calf. 20 mm.
IM.43882 (*Raschid* 27); IM 43900 (62).
Envelope.

150.
a. Sun God in ascending posture, placing one foot

on platform, and extending right arm. Two columns of inscription. Male figure wearing round cap and short kilt. 22 mm. (Pl. XXI-10)
b. The Weather God standing on his bull. Behind him, the hand of a worshipper or nude attendant holding a flowing vase. 20 mm. (Pl. VIII-2)
IM 43885 (*Raschid* 101).
Envelope.

151.
Hands of suppliant goddess behind worshipper offering kid. Between them porcupine? above, and fly below. 25 mm.
IM 43886 (*Raschid* 101).
Envelope.

152.
a. Impression in two registers. Above, worshipper facing god in ascending posture; winged lion demon. Below: lion attacking kneeling goat or bull; nude hero overpowering lion by holding it in reversed position. 16 mm.
b. Faint impression. Lion-headed demon. Human victim being trampled by a figure, possibly another demon? kneeling lion-headed demon. Star in sky. Notice that impressions are in reversed order. 22 mm.
IM 43897 (*Raschid* 44).
Envelope.

153.
Goddess Ishtar; back of figure with mace is preserved. 20 mm.
IM 43908 (*Raschid* 117).
Envelope.

6.4 Tell Ed-Der

154.
a. Fragment, damaged. Partial figure of god; kneeling nude hero with curls lifting lion over his head. Figure with mace facing god? who raises one hand. Box of inscription between them. 16 mm.
b. Part of Sun God holding saw in ascending posture. 20 mm.
c. Partial impression of figure with mace between two suppliant goddesses, the seal was originally capped. 15 mm. (Pl. XXXIII-2)
L. III; IM 51024; *Edzard* No. 35.
Envelope; contract; sale of field; Sabium 13th year.

155.
Badly damaged impression. Worshipper facing enthroned deity? Crescent in between them. Three drill holes above reclining goat. 15 mm.
L. II; Rm 14; IM 49301; *Edzard* No. 35.
Envelope; contract; sale of field; Sabium, 13th year.

156.
Fragments. Worshipper, head missing, with clasped hands facing enthroned "king" on fleeced stool over platform. In sky: sun disc in crescent. Terminal: two columns of inscription. 20 mm. (Pl. XXIV-4)
L. III; IM 49254; *Edzard* No. 155.
Envelope; account for bread, grain and flour.

157.
a. Suppliant goddess behind worshipper who raises one hand in supplication to enthroned god holding ring and staff. 17 mm.
b. Part of Sun God holding saw. Conquering God trampling on victim, only partly visible. 20 mm. (Pl. XXIII-2)
L. III; Rm.3; IM 49959.
Lump of clay.

158.
a. Partial impression, shows enthroned deity (stool missing), over bull on an upper register with a winged lion-dragon below. Line from bull's neck to standing figure with mace; this could be a break in the seal. 20 mm. (Pl. XXIII-7)
b. Lower part of an impression. Nude priest, holding pail and standing over double-tier red platform. Sun God holding saw and placing one foot on mountain. 15 mm.
c. Damaged surface. Part of nude female, and the goddess Ishtar. Unidentifiable figure behind. 15 mm.
d. Enthroned "deity" wearing round cap and holding cup, is flanked by two worshippers. Winged lion-headed demon may have been added later between the worshippers. The seal is in the Old Assyrian style. 10 mm. (Pl. XLIV-1)
e. Only a "Humbaba" head.
L. II; IM 48822; *Edzard* No. 2.
Envelope; loan of silver; Anum-pisha archive.

159.
a. Partial impression. God holding club? in ascending posture, facing left, probably toward group

consisting of priest with pail and sprinkler turned toward Weather God who holds lightning fork and stands on fire-spitting dragon. All facing right. 20 mm. (Pl. XXIII-8)

b. Four columns of inscription (fourth column mostly broken). Figure, head not preserved, wearing long plain garment, holds weapon in lowered hand. 20 mm.

c. God in ascending posture, partially preserved, turned to left, beside him partial nude attendant with curls who holds vase from which water flows. 19 mm.

d. Attenuated stick figure (secondary addition). Weather God standing on humped bull holds weapon over shoulder and lightning fork in other hand, and faces a suppliant goddess. 22 mm. (Pl. XXIII-9)
L. II; IM 50413; *Edzard* No. 21.
Envelope; loan of silver; Anum-pisha archive.

160.
a. Nude attendant with curls holds flowing vase. Suppliant goddess faces figure with mace. Second nude attendant with curls offering kid to goddess Ishtar. 15 mm.
b. The goddess Ishtar; worshipper wearing brimmed cap, offers kid to Sun God holding saw, and stepping on stylized mountain. 25 mm.
L. II; IM 48826; *Edzard* No. 24.
Envelope; loan of grain; Anum-pisha archive.

161.
Two nude bearded heroes with curls grasp reversed lion which seems to bite first hero's knee; goddess Ishtar stepping on stylized lion holds scimitar and double lion-headed club. Human victim, wearing round cap and short pointed beard, is attacked by lion-headed winged dragon. In field, severed human head and cuneiform sign "DINGIR" close to lion dragon. 22 mm.
L. II; IM 49320; *Edzard* No. 11.
Envelope; loan of silver; Anum-pisha archive.

162.
Suppliant goddess behind worshipper wearing brimmed cap and long fringed garment; offering kid to Sun God who holds saw and places one foot on stylized mountain. Behind him stands figure with mace wearing fan-shaped beard. Terminal: two columns of inscription. 20 mm.
L. I; IM 49265; *Edzard* No. 16.
Envelope; loan of silver and gold; contemporary Anum-pisha.

163.
Worshipper (possibly with offering), faces god in ascending posture. Secondary motif in two registers; above goat attacked by lion who in turn is being restrained by bull-man; below, two lions attacking goat; nude hero with curls grasps one lion. 25 mm. (Pl. XXIII-3)
L. III; IM 50427; *Edzard* No. 19.
Envelope; loan of silver; contemporary Anum-pisha.

164.
Bull-man and nude bearded attendant with curls hold triangle-topped standard between them. Suppliant goddess and worshipper wearing brimmed cap advance toward enthroned "king" sitting on fleecy stool holding cup. In field: vase above ball-staff; sun disc in crescent above dog balancing crook. 20 mm. (Pl. XXIV-2)
L. III; IM 49541; *Edzard* No. 28.
Envelope; loan contract; contemporary Anum-pisha.

165.
a. Worshipper in long garment and brimmed cap raising one hand in front of his face, stands opposite Sun God who holds saw and places one foot on stylized mountain; to right, figure with mace faces suppliant goddess (only lower part of latter is visible). 22 mm.
b. Figure with mace behind suppliant goddess who stands behind worshipper offering kid to god holding staff? Between them small figure of nude bearded attendant with pairs of curls, holding a standard? with tassel. 14 mm.
c. Partial impression of priest with pail and sprinkler facing god in ascending posture who holds saw in vertical position. Remains of third figure visible in photograph. 19 mm.
d. Partial impression showing god turned right; turned to left, bull-man. 14 mm.
e. Head of female worshipper? 12 mm.
L. II; IM 50898.
Envelope; contemporary Anum-pisha.

166.
a. Deity (head damaged), with one hand raised behind worshipper with clasped hands, who faces nude male figure. In field: crescent above crook; dog balancing staff. 16 mm. (Pl. XXIV-3)
b. Part of nude bearded figure with curls beside nude female. 12 mm.
c. Lower part of striding animal. Lower part of figure, possibly of worshipper. 10 mm.
d. Part of nude bearded figure with curls carrying flowing vase. 12 mm.
L. II; IM 50426; *Edzard* No. 33.
Envelope; purchase of field; contemporary Anum-pisha.

167.
First scene, at left: small male figure wearing round cap and girdle, stands on panelled platform, facing goddess holding ring and staff standing on platform supported by two seated dogs. In sky: sun disc in crescent. Behind goddess, small figure of nude bearded attendant with curls. Second scene, at right: suppliant goddess behind bareheaded worshipper facing enthroned "king," holding cup, and sitting on fleeced stool over platform. In field: sun disc in crescent above mongoose. 24 mm. (Pl. XXIV-1)
L. II; IM 49347.
Envelope; contemporary Anum-pisha.

168.
Suppliant goddess following worshipper who raises one hand forward before enthroned "king," who sits on fleecy stool. Sun disc in crescent in sky. Secondary scenes: above, two human-headed bulls are seated opposite each other, one human head is above bull at right; below, suppliant goddess faces nude? figure holding vessel? followed by nude bearded hero with curls. Surface of impression damaged. 18 mm. (Pl. XXIII-5)
L. II; IM 50437.
Envelope; contemporary Anum-pisha.

169.
Late Old Assyrian style seal impression in two registers. Above, two groups: in one, two worshippers approach enthroned "king," serpent behind throne; in second group, three more worshippers face bull, who stands on platform, ball-staff before him. – Separated by line. Below is one group of two figures approaching enthroned "king"; second group has a male figure to left of symmetrical group of two male figures, each holding one pair of human-headed bulls with heads turned backwards.
L. II; IM. 50471; found together with *Edzard* No. 183.
Bulla.

170.
a. Partial impression of suppliant goddess. 18 mm.
b. Sun God holding saw placing one foot on artificial mountain. 12 mm.
c. Part of presentation scene. Worshipper facing enthroned "king." 11 mm. (Pl. XXIV-7)
Envelope.

171.
Possibly impressions from several seals. Part of human victim; nude bearded hero with curls in contest with bull-man; upside down animal; suppliant goddess facing Weather God holding lightning fork turned to left. Second god turned to right partly obliterated; worshipper holding offering. Cuneiform signs scattered among the figures. 15 mm.
L. II; IM 50498; *Edzard* No. 37.
Envelope; contract of field purchase; Samsuiluna, 18th year.

172.
a. Suppliant goddess facing columns of inscription. 16 mm.
b. Impression in the drill style. God in ascending posture, brandishing weapon; second god possibly facing another deity who holds stylized lion club. 16 mm. (Pl. XXXVI-5)
L. I; IM 50868; *Edzard* No. 45.
Tablet; receipt; Ammiṣaduqa, 2nd year.

173.
Drilled style impression. Vanquishing god in short mantle, trampling enemy. God in long garment turning right, away from contest group. Drill holes in field. 15 mm. (Pl. XXXVII-4)
L. I; IM 50894.
Tablet; found with *Edzard* No. 31; dated to Ammiṣaduqa, 2nd or 4th year.

174.
Drilled style. Part of ascending god turned left. Conquering god turned right. Part of enthroned god holding staff turned left. Inscription belongs to another seal. 18 mm.
L. I; IM 50893. *Edzard* No. 46.
Tablet; receipt for silver; Ammiṣaduqa, 2nd or 4th year.

175.
Syrian seal. Partly impressed over inscription of another seal. Figure with raised hand, wearing fringed garment, which winds round body, faces nude goddess. Between them small figure of worshipper. Behind goddess is figure with raised hands, possibly suppliant goddess. In field: sun disc made with drill holes, ball-staff; guilloche as upper border. 15 mm. (Pl. XXXVII-12)
L. I; IM 50892.
Tablet; found with No. 21; Ammiṣaduqa, 2nd or 4th year.

176.
Faint impression. Figure with one hand raised facing god in ascending posture holding globe standard. Enthroned god holding crook. Inscription belongs to another seal. 15 mm.
L. II; IM 50418.
Tablet; loan contract; Ammiṣaduqa, 7th year.

177.
Figure with mace, facing suppliant goddess; between them sun disc executed with drill holes. 15 mm.
L. II; IM 50418.
Loan contract; Tablet; Ammiṣaduqa, 7th year.

178.
Impression obliterated by text of tablet. Devotional scene. Part of suppliant goddess, worshipper and enthroned deity. 15 mm.
IM 50421; *Edzard* No. 39.
Tablet; Ammiṣaduqa, 7th year.

179.
Three impressions in drill style.
a. Possibly suppliant goddess with raised hands. Nude bearded hero with curls. Disc in sky executed with drill holes. 14 mm. (Pl. XXXVI-8)
b. Enthroned god holding emblem partly damaged. Figure executed with drill and deep vertical incisions. 14 mm. (Pl. XXXVI-9)
c. Weather God standing on bull. 16 mm. (Pl. XXXVII-6)
L. I; IM 50869; *Edzard* No. 48.
Tablet; receipt; Ammiṣaduqa, 11th year.

180.
Weather God in drill style, standing on bull and holding its rein. 20 mm. (Pl. XXXVII-7)
L. II; IM 50419; *Edzard* No. 50.
Tablet; loan of silver; Ammiṣaduqa, 14th year.

181.
Several impressions in drill style.
a. Figure with one raised hand facing second figure holding staff; conquering god at right. 14 mm.
b. Suppliant goddess facing god? holding symbol (damaged), between them nude female with head in profile. 14 mm.
c. Suppliant goddess facing god? holding symbol (damaged). Between them female with head in profile. 14 mm.
L. II; IM 50432; *Edzard* No. 41.
Tablet; field contract; Ammiṣaduqa.

6.5. Sippar

182.
a. Enthroned "king" facing left holding cup, seated on fleece-covered stool placed on platform. Three columns of inscription. Figure with mace facing right. In sky: disc in crescent. 25 mm. (Pl. XXIV-5)
b. Goddess Ishtar, placing one foot on lion, holding multiple weapon of four maces and three double lion-headed club-maces. On either side of her head, one spread-winged eagle. Scene of presentation: suppliant goddess behind bareheaded worshipper, facing enthroned "king" holding cup sitting on fleecy stool over platform. In sky: star disc in crescent. 23 mm. (Pl. XXII-1)
c. Suppliant goddess, behind worshipper with raised hand carrying kid as offering to enthroned god, who sits on bull-shaped throne placed on platform. Terminal motif: nude bearded hero with curls, in profile, attacking lion. In field: scattered cuneiform signs; star disc in crescent in sky. 22 mm. (Pl. XXII-2)
d. Sun God holding saw vertically, enthroned on high-backed seat facing left; two columns of inscription; partial impression of suppliant goddess facing right. 21 mm. (Pl. XXII-3)
e. Conquering king in tall cap, brandishing weapon and conquering enemy, helped by warrior wearing round, slightly pointed hat, carrying bow over shoulder. Above victim, seated goat with head turned back. Secondary motif in two registers: above, two lions attacking human victim; below, two crossed lions each attacking goat. Two drillings in field. 25 mm. (Pl. XXII-4)
CT Vol. 45, No. 1.
Envelope; contract; Buntahtunila.

183.
Conquering king brandishing weapon and trampling enemy. Goddess Ishtar holding weapons. By her shoulder small kneeling human figure, unidentifiable object, possibly a bird; below, captive lion-headed demon. God with bird foot, trampling human victim and thrusting weapon at him. Lion-headed demon, holding with one hand double lion-headed club-mace and human victim with other. 20 mm. (Pl. XXII-5)
Louvre II; A. 477; Pl. 112, Fig. 9. *ThD*, No. 56.
Envelope; contract; Sumulael.

184.
a. Suppliant goddess behind worshipper facing enthroned "king" holding cup. Heads either sketchy or missing. 19 mm. (Pl. XXIV-6)
b. Worshipper wearing round cap with upturned brim

offering kid to god, figure with mace facing suppliant goddess. 15 mm.
CT 47, No. 2a.
Envelope; Apil-Sin.

185.
a. Turned to left, god in ascending posture resting hand on scimitar. Turned to right, suppliant goddess behind worshipper offering kid. In sky: human head; disc in crescent. 17 mm.
b. God holding with left hand small human victim (possibly this is the same god who stands at end of scene, holding with right hand what appears to be a fish); conquering god trampling on human victim. Lion-headed demon holding animal?, following suppliant goddess who faces god with flowing vase, whose water is received by kneeling small man holding up vase. In sky: vase, four drill holes. 22 mm. (Pl. XXIII-1)
c. Upper part of bull-man, nude bearded hero with curls carrying kid offering; between them two reclining animals, facing each other above two small male figures whose heads are preserved. 16 mm. (Pl. XXIII-6)
d. Suppliant goddess facing goddess Ishtar holding her weapons and stepping on lion, fish between them; worshipper offering kid to Sun God who places one foot on artificial mountain. In sky: hedgehog; disc in crescent. 23 mm.
CT 47, No. 3a.
Envelope; sale of field; Apil-Sin.

186.
a. Enthroned "king" on fleecy stool. 19 mm. (Pl. XXIV-8)
b. Worshipper (lower part visible). Sun God holding saw and placing one foot on artificial mountain; four columns of inscription. 23 mm.
c. Female worshipper with top hat (priestess); four columns of inscription. Seal was possibly capped. 2 mm.
CT 47, No. 7a.
Envelope; adoption document; Apil-Sin.

187.
a. Worshipper, one hand raised, offers kid to Sun God. 22 mm.
b. Two columns of inscription visible. Figure with mace. 23 mm.
c. Three columns of inscription; suppliant goddess. 24 mm.
d. & e. Figure wearing short tunic with point between legs holding down scimitar, behind worshipper, who wears long plain garment and round cap marked with parallel lines and faces god (partly visible), who places one foot on platform and holds ring. In sky: disc in crescent. 17 mm.
f. Bull-man in contest with nude hero. 22 mm. (Pl. XXX-1)
CT 47, No. 10a; *Harris*, pp. 222, 229.
Envelope; lease of field; Sin-muballiṭ, 8th year.

188.
a. Four columns of inscription; partial impression of god with mace. 22 mm.
b. Three columns of inscription; nude bearded hero with curls. Drill holes prominent on shoulders and thighs. 24 mm.
c. Turned to left, suppliant goddess, followed by female worshipper, three columns of inscription. 23 mm. (Pl. XXXIII-5)
d. Two registers divided by horizontal line. Probably secondary motif of a cylinder. Above, worshipper between two suppliant goddesses. Below, worshipper? behind second worshipper who offers kid to god in ascending posture. 20 mm.
e. Worshipper with clasped hands behind female worshipper facing enthroned god sitting on panelled stool holding ring and staff. In field: vase above ball-staff, reclining animal, monstrous mask, star disc in crescent above nude priest with pail and sprinkler. 22 mm. (Pl. XXV-5)
f. Partial impression of god. 21 mm.
CT 47, 21a; *Harris*, pp. 61, 129, 191, 261, 313.
Envelope; Sin-muballiṭ, 9th year.

189.
a. Old Assyrian in two registers. Above, deity behind worshipper; cuneiform signs. Below, human victim attacked by two lions. 21 mm. (Pl. XLIV-2)
b. Suppliant goddess turned to left, three columns of inscription; bull-man in contest above nude hero in contest (partners missing). 21 mm. (Pl. XXXIII-4)
c. God in ascending posture. 21 mm.
d. Worshipper offering kid to Sun God holding saw and placing one foot on artificial mountain. In sky: fly, bull, severed human head. 22 mm. (Pl. XXV-2)
e. Damaged impression; part of inscription. Upper part of two gods facing each other; nude female. In field: vase above ball-staff. 14 mm.
f. Old Assyrian impression in two registers. Above: lightning fork above vase, worshipper, winged lion dragon. Below: bow-legged dwarf, two crossed lions. 21 mm. (Pl. XLIV-4)
g. Old Assyrian impression; lion, bull-man fighting lion. 22 mm. (Pl. XLIV-5)
Louvre II, A. 517; *ThD* No. 74.
Envelope; Sin-muballiṭ.

190.
a. Two columns of inscription, part of worshipper. 23 mm.

b. Two columns of inscription, part of figure with mace. 27 mm.
c. Three columns of inscription, figure with mace. 25 mm.
d. God holding ring and staff seated on panelled throne placed on platform. Beside throne is full-faced goddess with clasped hands. In sky: fish-man, disc in crescent. 24 mm.
e. Full-faced goddess with clasped hands beside figure in tall cap and flounced robe offering kid to goddess Ishtar who stands on two lions and holds double lion-headed club. Cuneiform signs scattered in field. 24 mm. (Pl. XXVII-5)
f. Nude figure with round cap holding standard, following worshipper, wearing round cap and long plain garment, one hand raised facing damaged figure. 15 mm.
g. Bull-man offering kid to god enthroned on panelled stool over platform, holding ring and staff. Behind is full-faced goddess raising one hand. In field: goat above tortoise, disc in crescent above bow-legged dwarf. 25 mm. (Pl. XXVII-6)
h. "King" sitting on cross-legged stool. 20 mm.
i. Full-faced goddess with one hand raised. In sky: goat-fish. Bearded god with mace, full-faced goddess with clasped hands; between them, cuneiform inscription. 24 mm. (Pl. XXVII-1)
CT 47, No. 11a; *Harris*, pp. 157, 194, 218, 380.
Envelope; sale of field; Sin-muballiṭ.

191.
a. Part of an enthroned god, back of standing god. 16 mm.
b. Part of goddess Ishtar. Three columns of inscription. Suppliant goddess. 21 mm.
CT 47, No. 13a; *Harris*, pp. 218, 298.
Envelope; sale of field; Sin-muballiṭ.

192.
a. Human-headed fish above hedgehog above nude female; three columns of inscription; lower part of goddess. 24 mm.
b. Three columns of inscription; part of figure with mace. Lightning fork in sky. Same impression as No. 196. 25 mm.
c. Three columns of inscription; part of god with mace. 18 mm.
d. Worshipper with one hand raised, offers kid to enthroned deity (larger part missing) who rests feet on reclining bull; disc in crescent in sky. In field: mongoose. 21 mm.
e. Back of god; part of inscription. 22 mm.
f. Back of god? Three columns of inscription; back of suppliant goddess. 25 mm.
CT 47, No. 15a; *Harris*, p. 26.
Envelope; sale of house; Sin-muballiṭ.

193.
a. Suppliant goddess. 19 mm.
b. Partial impression of goddess Ishtar, stepping on her lion and holding double lion-headed club-mace. 19 mm.
c. Worshipper offering kid. In sky: disc in crescent. 20 mm.
CT 47, No. 17a; *Harris*, p. 33.
Envelope; sale of land; Sin-muballiṭ.

194.
a. Two columns of inscription. Worshipper offering kid to god (Shamash?). In sky: disc in crescent. 21 mm. (Pl. XXXII-5)
b. Worshipper with clasped hands, wearing long garment and round cap with parallel lines, facing enthroned "king" on fleece-covered stool. In field: crescent above bow-legged dwarf; cuneiform signs, small male nude figure and crouching dog balancing crook. 20 mm.
c. Part of enthroned "king" on fleece-covered stool facing left; facing right, bull-man fighting with lion; part of suppliant goddess. 26 mm.
d. Figure with mace facing god? who raises one hand, between them crescent above nude female. 17 mm.
e. Three columns of inscription, figure with mace facing suppliant goddess, followed by the god Amurru. 21 mm.
CT 47, No. 19a.
Envelope; Sin-muballiṭ.

195.
a. Part of (suppliant) goddess turned to left; nude hero in contest with bull-man; between them severed human head. Nude female, "god" with mace turned to right. 20 mm. (Pl. XXX-2)
b. Full-face bearded god with clasped hands; goddess full-face; part of nude hero. In field: monstrous head, star disc in crescent, cuneiform signs and severed human head. 18 mm. (Pl. XXXII-1)
c. Goddess Ishtar, suppliant goddess turned to left. 21 mm.
e. Partial impression of full-face goddess; figure with mace, disc in crescent above nude female. 20 mm.
f. Part of enthroned deity holding cup. In sky: disc in crescent. 19 mm.
g. Suppliant goddess, figure with mace who faces second suppliant goddess. Between them star disc in crescent above nude female. 22 mm. (Pl. XXXIII-1)
h. Part of goddess Ishtar; four rows of inscription, two of them inside columns; bearded hero with triple curls; suppliant goddess. 24 mm. (Pl. XXIX-3)
x. God? in flounced garment offering kid. 15 mm.
Louvre II, A.518; *ThD* No. 73.
Envelope; Sin-muballiṭ.

196.
a. Goddess Ishtar, three columns of inscription; god with mace. 25 mm.
b. Three columns of inscription; figure with mace; lightning fork in sky. Same impression as No. 192.b. 20 mm.
c. Figure with mace on double tiered platform. 23 mm.
d. Nude priest with pail and sprinkler on platform; facing god? who wears short garment opening down front. 20 mm. (Pl. XXXI-6)
e. Suppliant goddess facing inscription. 21 mm.
Louvre II, A.520; *ThD* 76.
Envelope; Sin-muballiṭ.

197.
a. Suppliant goddess following worshipper offering kid to Sun God; disc in crescent sky. Bull-man in contest with nude hero. Between them: seated bull above lion-fish. 17 mm. (Pl. XXX-3)
b. Old Assyrian impression. Part of supplian goddess behind worshipper facing god with crescent on his crown, placing one foot on artificial mountain and holding mace. Behind him nude hero with curls in profile in fighting posture. In field: vase above ball-staff, scorpion. 19 mm. (Pl. XLIV-6)
c. Three columns of inscription. Worshipper, bull-man holding sun disc standard. Sun God places foot on artificial mountain. Lightning fork in field. 22 mm. (Pl. XXXII-6)
d. Goddess Ishtar in profile ascending, holding stylized double lion-headed mace-club. 21 mm.
e. Nude priest with pail and sprinkler. 13 mm.
f. Three columns of inscription, back of figure with mace. 16 mm.
Louvre II, A.520; *ThD* 76.
Envelope; Sin-muballiṭ.

198.
a. Impression of Agade seal. Part of nude bearded hero with short hair; box of inscription above animal. Nude hero with curls in contest. 24 mm.
b. Nude bearded hero turned to left; suppliant goddess. Star disc in crescent, back of mongoose. 27 mm. (Pl. XXX-4)
c. Figure with mace; front of seated goat above unidentifiable object above back of rampant lion. 25 mm.
d. Sun God. Two small figures of nude priests with pail and sprinkler, one above the other. 24 mm.
e. Part of goddess Ishtar; nude bearded hero with clasped hands. God holding vase with single stream of water flowing down into second vase. 24 mm. (Pl. XXVI-4)
f. Figure with mace facing suppliant goddess; between them seated bull above nude female. Full-face water goddess standing on two goat-fish; fish swim up by her sides, human-headed fish by her head; beside her, full-face goddess. 25 mm. (Pl. XXVI-7)
g. God in flounced garment offering kid to second god, disc in crescent. 27 mm.
h. Figure with mace facing goddess Ishtar? Severed human head in sky. Seal was capped. 20 mm.
i. Full-face goddess Ishtar holding double lion-headed club, standing on two lions. Suppliant goddess. Nude female and seated bull as fill. 22 mm. (Pl. XXV-7)
j. Peripheral seal, possibly Elamite. Three columns for inscription, suppliant goddess behind worshipper who extends both hands toward enthroned king. 18 mm. (Pl. XLIV-7)
Louvre II, A.522, Pl. 114, Fig. 1; *ThD* No. 75.
Envelope; Sin-muballiṭ.

199.
a. Bearded hero with curls. God? with clasped hands. Bull-eared god holding down scimitar and stepping on lion-headed demon who brandishes knife. In sky: human-faced reclining bull. Figure with mace facing suppliant goddess. 25 mm. (Pl. XXIX-7)
b. Figure with mace, facing suppliant goddess. Between them, star disc in crescent. 22 mm.
Louvre II, A.523, Pl. 114, Fig. 3; *ThD* No. 79.
Envelope; Hammurabi, 2nd year.

200.
a. & b. It is difficult to separate the two impressions. Upper part of nude bearded hero with curls; full-faced water goddess; suppliant goddess facing full-faced goddess; human victim being overpowered by lion-headed winged dragon; suppliant goddess behind figure with mace; between them, a fly. 24 mm. (Pl. XXXI-5)
CT 47, No. 25a; *Harris*, pp. 245, 301, 327.
Envelope; sale of field; Hammurabi, 3rd year.

201.
a., b. & c. Goddess with clasped hands facing full-face goddess (Ishtar), standing on two lions, holding club (not preserved), above seated goat. Suppliant goddess behind worshipper offering kid to Sun God who holds saw and places one foot on artificial mountain; behind latter, nude bearded hero with curls holding vase. In field: hedgehog above lion-fish. In sky: bull. 22 mm. (Pl. XXV-8)
d. & e. Deity with clasped hands facing Sun God holding saw, placing one foot on artificial mountain. Figure with mace facing suppliant goddess. In sky: fly, goat sitting on knoll (missing), reclining goat with head turned backward. In field:

five drilled globes, crook above nude female, lion- -fish. 24 mm. (Pl. XXV-4)

f. Nude hero with curls fighting a lion, between them, head of bull-man. Second nude hero fighting (animal missing). In sky: cuneiform sign. 24 mm. (Pl. XXX-5)

g. Three columns of inscription. God with mace. 20 mm.

h. Enthroned god on high-backed chair decorated with scale pattern. Bearded worshipper embracing goddess (only small part visible). Cuneiform signs in field. 20 mm. (Pl. XXV-1)

i. Worshipper offering kid to god holding ring and placing one foot on artificial mountain. In sky: sun disc in crescent. 17 mm.
CT 47, No. 27a; *Harris*, pp. 245, 261.
Envelope; sale of threshing floor; Hammurabi, 4th year.

202.

a. Three columns of inscription, worshipper wearing long garment and round cap with upturned brim. 26 mm.

b. Enthroned god facing left. Behind throne: minor god turned to left, with clasped hands; back portion of god with mace to right. In sky: reclining bull, head of nude bearded hero with curls; nude female below. 25 mm. (Pl. XXVII-2)

c. Contest scene. Lion-griffin attacking kneeling human victim, nude bearded hero with curls, in profile, grasping lion, who attacks second hero who in turn overpowers lion, holding him up by hind leg. 20 mm. (Pl. XXXIII-7)
CT 47, No. 28a; *Harris*, pp. 245, 249.
Envelope; sale of threshing floor; Hammurabi, 7th year.

203.

a. Three columns of inscription. Back of suppliant goddess. 22 mm.

b. Back of god turned to left. Terminal in two registers: cow suckling calf, dividing line. Below, goat? and nude male figure holding standard. Figure with mace turned to right. 23 mm.

c. God (Shamash) to left, placing one foot on artificial mountain. Nude bearded hero with curls fighting lion (head missing). Back of goddess to right. 22 mm. (Pl. XXX-6)

d. Suppliant goddess behind god offering kid to Sun God. 22 mm.

e. Full-face goddess; beside her, suppliant goddess to left. 19 mm. In sky: sun disc in crescent.
CT 47, No. 29a; *Harris*, pp. 26, 299.
Envelope; sale of house; Hammurabi, 7th year.

204.

a. Bull-man offering kid to Shamash. Goddess full-face; back of deity turned to left. In sky: sun disc; monstrous mask below, part of animal. 22 mm.

b. Nude hero, grasping lion in inverted position. Rampant lion. 23 mm.
CT 45, No. 20; *Harris*, p. 274.
Envelope; exchange contract; Hammurabi, 9th year.

205.

a. Figure with mace turned to right; upper part of small nude female. 18 mm.

b. Embracing couple (goddess and worshipper). Suppliant goddess turned to right. 22 mm. (Pl. XXXII-3)
CT 47, No. 30a; *Harris*, pp. 157, 164.
Envelope; dowry; Hammurabi, 10th year.

206.

a. God turned to left; three columns of inscription; back of figure with mace turned to right. 20 mm. (Pl. XXIX-2)

b. God? with mace facing suppliant goddess, between them, figure of small nude male. In sky: sun disc in crescent. Three columns of inscription. 24 mm. (Pl. XXXIII-3)

c. Four columns of inscription; nude bearded hero with curls offering kid to enthroned Water God, who holds vase, water flowing down from his elbow. God rests feet on goat fish; another goat-fish supports panelled throne. Behind throne, full-faced goddess. In sky: sun disc in crescent, tortoise. 22 mm. (Pl. XXVI-1)

d. God with mace, facing nude female. God facing large lion-headed giant club. 20 mm.

e. Nude hero in contest with bull-man, worshipper offering kid to Sun God. Heads of all figures obliterated. 20 mm. (Pl. XXX-7)

f. Deity in flounced garment facing Sun God. Between them: reclining humped bull; below, monstrous mask. 24 mm. (Pl. XXV-3)
Louvre II, A.524, Pl. 114, Fig. 4; *ThD* No. 83.
Envelope; Hammurabi, 10th year.

207.

a. Nude hero with curls in profile to left grasping lion (only back is visible). Terminal motif: figure with mace divided by horizontal line from nude priest with pail and sprinkler below. Suppliant goddess behind worshipper holding kid, only front part is visible. 16 mm. (Pl. XXXI-7)

b. Face and hands of suppliant goddess; goddess full-face. 29 mm.
CT 47, No. 31a; *Harris*, pp. 8, 61, 75, 121.
Envelope; law suit; Hammurabi, 11th year.

208.
a. Deity (head missing), hands folded to its chest; curly haired nude hero kneeling on one leg, overpowering lion by holding it over its head. In field: crook above seated animal (head missing); cuneiform signs. 25 mm.
b. God with mace facing suppliant goddess. In sky between them: seated reclining lion. 23 mm. (Pl. XXVII-4)
c. God holding hook attached to fish. 32 mm. (Pl. XXXII-4)
d. & e. Suppliant goddess behind god offering kid. In sky: cuneiform signs, star disc in crescent, seated animal. 25 mm. (Pl. XXX-9)
CT 47, No. 32a; *Harris*, pp. 218, 377.

209.
a. Full-face goddess. Short-kilted conquering god brandishing scimitar with right hand, with left multiple weapon of seven double lion-headed clubs, while trampling enemy. Nude bearded hero with curls holding vase. In sky: reclining human-faced bull. 20 mm. (Pl. XXVII-3)
b. Four columns of inscription; figure with mace, facing suppliant goddess. 25 mm.
c. Syrian seal. Conquering king in chariot, presented by goddess in open skirt with spear and curved weapon. Above chariot, snake goddess (head missing) engraved horizontally. Terminal in two registers separated by guilloche: upper, lion attacking goat; lower, row of five warriors. 13 mm. (Pl. XXXVII-11)
d. Two columns of inscription visible; suppliant goddess behind worshipper who offers kid to deity of which only hand holding staff is visible. 20 mm.
e. Back of figure to left, nude beareded hero with curls offering kid to goddess Ishtar. In sky: sun disc in crescent. 22 mm.
CT 47, No. 22a; *Harris*, pp. 33, 299.

210.
a. God to left, nude female, upside down; figure with mace. Fly in sky. In field: spread-winged eagle above seated animal. 16 mm.
b. & c. Worshipper offering kid to god enthroned on panelled seat placed on platform decorated by triangles. Behind throne is full-face goddess. Second full-face goddess, lacking side curls, is beside first, with third suppliant goddess with raised hands. 22 mm. (Pl. XXVII-7)
d. God with clasped hands wearing bordered garment, facing god in ascending posture. Beyond, worshipper. Fly in sky. 23 mm.
e. Legs of figure; front of face and arm of another figure holding standard; giant lightning fork and lion-headed club, nude female, legs of figure. 16 mm.
f. Back of deity, to left upper part of god with mace. In field: front of seated bull, above small figure of worshipper. 20 mm.
CT 47, No. 23a; *Harris*, pp. 325,326.

211.
a. Water God holding flowing vase, standing on goat-fish. Lightning fork, separated by horizontal line from nude female below. Nude bearded hero with curls in contest with bull-man; suppliant goddess turned to right. 25 mm. (Pl. XXVI-3)
b. Part of bull-eared god in "sarcophagus," miniature suppliant goddess; three columns of inscription; figure with mace facing suppliant goddess, only one arm is visible. 22 mm.
c. Figure with mace, facing suppliant goddess; between them; crescent standard, below double lion-headed club-mace; snake; spade. 22 mm.
d. & e. Deity (part of flounced garment is visible); miniature nude female; worshipper offering kid to god in ascending posture, holding serrated ring. Bull-man in contest with nude bearded hero with curls, between them seated goat. 22 mm. (Pl. XXX-8)
f. & g. God with mace facing suppliant goddess; between them: cuneiform signs. Goddess behind worshipper who offers kid to god enthroned on platform who holds staff. In sky: star disc in crescent, severed human head. 19 mm.
CT 47, No. 34a; *Harris*, pp. 33, 299, 326.

212.
a. & b. Bull-man in contest with nude hero (missing); between them: lion fish. Suppliant goddess accompanying god offering kid to goddess (probably Ishtar); between them: cow suckling calf. In field: nude priest with pail and sprinkler, above animal (dog?); severed human head, above small nude female. 22 mm.
c. Back of enthroned god turned to left, full-face goddess. Between them: reclining bull, cuneiform inscription. 19 mm.
d. Suppliant goddess beside full-face goddess with clasped hands accompanying worshipper offering kid to enthroned Water God on platform which consists of two goat-fish. Behind throne, full-face goddess. In sky: goat-fish, sun disc in crescent. 19 mm. (Pl. XXVI-6)
e. Full-face goddess beside bearded god offering kid to enthroned god on platform, holding staff, behind throne, suppliant goddess (only hands visible). Severed human head and goat in sky. 22 mm. (Pl. XXVII-6)
f. Frontal-face goddess; goddess with clasped hands approaching enthroned god holding staff. In sky: bull. 21 mm.

CT 47, No. 30a; *Harris*, pp. 72, 219, 259.
Envelope; sale of field; Hammurabi, 25th year.

213.
a. Back of an enthroned god. 18 mm.
b. Four columns of inscription; god with mace. 25 mm.
c. Three columns of inscription, figure with mace. 15 mm.
CT 47, No. 47a; *Harris*, pp. 29, 157, 198.

214.
a. Deity? offering kid to enthroned god holding ring and staff. 18 mm.
b. Lion attacking bull (only heads visible); small goat sitting on knoll (missing), back of rampant lion. 10 mm.
Louvre II, A.525. *ThD* No. 88.
Envelope; Hammurabi, 26th year.

215.
a. Three columns of inscription; figure with mace. 17 mm.
b. Three columns of inscription; god with mace. 25 mm.
CT 47, No. 42a; *Harris*, p. 158.
Envelope; adoption document; Hammurabi, 32nd year.

216.
a. God with mace, wearing long swinging beard and facing suppliant goddess (between them: goat-fish above seated goat); god wearing plain long-fringed garment and square-topped single-horned crown, with clasped hands, accompanying worshipper offering kid to enthroned god, holding ring and staff, enthroned on platform. Between them: cow suckling calf, above seated lion; severed human head behind crown of god. 24 mm. (Pl. XXVIII-5)
b. Back of god to left. Nude bearded hero with curls holding vase toward enthroned Water God, who holds vase and rests his feet on goat-fish. In field: head of bull-man above goat on knoll; cow suckling calf above goat-fish. 22 mm. (Pl. XXVI-2)
c. Nude hero? Two registers of terminal motif: above, lions attacking goat. Beneath goat, two reclining goats with heads turned backward. Lower register has two lions, each attacking goat. 20 mm. (Pl. XXIII-4)
d. Back of god to left. Full-face goddess, with clasped hands. Suppliant goddess facing goddess Ishtar standing on two lions (one obliterated), holding double lion-headed mace-club; between them: box of inscription. Behind goddess: scattered cuneiform inscription. Worshipper, turned to his left, is carrying kid offering. In sky: severed human head, head of nude bearded hero with curls, head of bull-man. 22 mm. (Pl. XXV-9)
e. Back of enthroned god to left. Behind throne: nude bearded hero with curls extending vase (obliterated). Goddess wearing single-horned crown, with clasped hands beside suppliant goddess who accompanies worshipper carrying kid offering. In sky: severed human head, goat-fish. 21 mm. (Pl. XXXI-2)
f. Worshipper offering kid to god in ascending posture holding staff; behind him, full-face goddess. Figure with mace turned to left. In sky: sun disc in crescent. 20 mm.
g. Part of goddess; worshipper saluting goddess Ishtar in ascending posture, holding weapons. Beside Ishtar, god with clasped hands wearing plain long garment standing to right, behind figure (only back part visible). 18 mm.
h. Front of figure with mace turned to right, facing suppliant goddess; between them: lion-fish. Lion held by nude bearded hero with curls in profile. Back of goddess to right. In sky: severed human head, hedgehog. 18 mm.
i. Suppliant goddess behind goddess with clasped hands who accompanies "god" offering kid to enthroned god on platform, holding staff; behind him: deity holding vase with plant. In sky: sun disc in crescent (similar god with plant: *Corpus*, No. 391; *VR* 503). 20 mm. (Pl. XXVIII-8)
j. Goddess facing inscription (part of one column is visible). 6 mm. (Pl. XXXII-8)
k. Two suppliant goddesses, one on either side of three columns for inscription. 6 mm. (Pl. XXXII-7).
Louvre II, A.527, Pl. 115, Fig. 1; *ThD* No. 99.
Envelope; Hammurabi, 35th year.

217.
Full-face goddess. Suppliant goddess accompanying male figure, wearing chignon of gods but plain garment, saluting Sun God. 22 mm. (Pl. XXIX-1)
Louvre II, A.528; *ThD* No. 101.
Envelope; Hammurabi, 35th year.

218.
a. Deity with one hand raised, the other covered by garment; full-face goddess; god holding serrated ring. Beside latter, suppliant goddess (only front preserved). In sky: two crossed bulls? 21 mm. (Pl. XXIX-4)
b. Three columns of inscription. Crook above small nude female. 23 mm.
IM 43487; (*Raschid* 135).
Envelope; Hammurabi, 43rd year.

219.

a. One column of inscription visible; suppliant goddess facing nude female (both heads obliterated). Between fingers, dog balancing crook. 15 mm.
b. Figure with mace facing suppliant goddess; between them, sun disc, mongoose below. Weather God placing one foot on artificial mountain and holding lightning fork; small nude female above bull, possibly standing on it. 17 mm.
CT 47, No. 35a.
Envelope; Hammurabi.

220.

Suppliant goddess behind worshipper offering kid to Sun God; between them, lightning fork above, bull below. To the right: bull-eared god in "sarcophagus." 15 mm. (Pl. XXXI-3)
CT 47, No. 37a.
Envelope; Hammurabi.

221.

a. Three columns of inscription; figure with mace wearing long swinging beard. In field: fish above spear. 25 mm.
b. & c. Nude bearded hero with curls (head missing), holding flowing vase; figure with mace (head missing). In sky: star disc in crescent, seated bull?, small nude female below. 15 mm.
d. Lion griffin in contest with a second?. Between them: small nude female. Figure with mace. Sun disc in crescent above bow-legged dwarf above male figure which appears to carry crook over shoulder. 20 mm. (Pl. XXXI-9)
e. Figure with mace, wearing long swinging beard, facing full-face goddess who extends one hand; between them: sun disc above small kneeling nude attendant. Worshipper raising one hand, facing god with clasped hands; between them: severed human head, seated goat below. Terminal: scattered cuneiform signs. 19 mm. (Pl. XXVIII-2)
f. God (head obliterated), in sarcophagus, inverted. Two short-kilted worshippers, one holding crook; between them: nude female. In field: fish, second fish? below: fly, human head wearing round cap. 20 mm. (Pl. XLIV-8)
CT 47, No. 40a; Biggs, *JNES* 28, 1969, p. 134.
Envelope; adoption document; Hammurabi.

222.

a. Legs of nude figure to left followed by small nude figure, lower part of suppliant goddess accompanying worshipper who offers kid to Sun God (only saw and foot on artificial mountain are visible). In field: small worshipper, vase. 18 mm.
b. Bareheaded worshipper led by interceding goddess; between them: vase above ball-staff. 22 mm.
c. Three columns of inscription; figure with mace wearing long "swinging" beard. 23 mm.
d. Enthroned "king" to left; figure with mace facing female figure. 20 mm. (Pl. XXXIII-6)
e. Bareheaded worshipper with clasped hands, facing enthroned deity on platform (only part of garment and platform are visible). 23 mm.
f. God with clasped hands accompanying worshipper who raises one hand in salutation to Sun God. In field: porcupine, mongoose below; sun disc in crescent, fly below. 22 mm. (Pl. XXXI-4)
g. Worshipper (lower part visible), facing enthroned god holding ring and staff. In sky: sun disc in crescent. 20 mm.
CT 47, No. 50a; *Harris*, p. 299.
Envelope; Hammurabi.

223.

a. Small figure of nude bearded hero with curls; three columns of inscription; god with mace; above shoulder lion-fish (head missing). 20 mm.
b. & c. Nude figure (?) wearing triple girdle carrying offering behind figure with mace, facing suppliant goddess; between them: sun disc in crescent, mongoose below. Worshipper with one hand raised, facing deity with clasped hands (head missing). 22 mm.
d. Sun God (head obliterated) turned to left; enthroned god on platform (lower edge only is visible). 22 mm.
e. Worshipper facing god holding ring. Behind him nude bearded hero with curls holding vase with water flowing to a second vase at the base. 15 mm.
CT 47, No. 51a; *Harris*, pp. 230, 299, 275.
Envelope; lease of field; Samsuiluna, 1st year.

224.

a. Weather God, holding lightning fork and placing one foot on kneeling bull, three columns of inscription. 25 mm. (Pl. XXXIV-8)
b. Enthroned god turned to left. Lion-griffin fighting with bull-man. 19 mm.
c. Part of full-face deity; god offering kid to Sun God (saw obliterated); lion-fish behind latter. In sky: cuneiform sign, cow suckling calf. 20 mm. (Pl. XXVIII-7)
d. Columns of inscription; figure with mace facing suppliant goddess. 24 mm.
CT 47, No. 52; *Harris*, pp. 299, 341-342.
Envelope; purchase of slave; Samsuiluna, 1st year.

225.

a. Front of suppliant goddess beside god, who faces second god who holds staff; between them: sun disc in crescent above fly. Bull-man in contest

(partner missing). 15 mm.
b. Three columns of inscription; back of figure with mace. 14 mm.
Louvre II, A.531, Pl. 115, Fig. 4; *ThD* No. 106.
Envelope; Samsuiluna, 2nd year.

226.
a. Several columns of inscription; upper part of figure with mace wearing long "swinging" beard. 7 mm.
b. Suppliant goddess behind full-face goddess, who stands beside worshipper wearing flounced garment of the gods, and offering kid to god of whom only hand holding ring and staff are visible. Cuneiform signs scattered in field. 27 mm. (Pl. XXXVIII-4)
c. Back? of god, nude bearded hero with curls. 14 mm. (Pl. XXXIV-3)
d. Hands of suppliant goddess behind worshipper carrying an offering. In sky: star disc in crescent. 21 mm.
CT 47, No. 56a; *Harris*, pp. 63, 219, 380.
Envelope; sale of field; Samsuiluna, 4th year.

227.
a. Three columns of inscription; back of god with mace. 21 mm.
b. Three columns of inscription; lower part of figure with mace. 20 mm.
c. God turned to left. Bull-man in contest (partner missing). 22 mm.
d. Bull-man with clasped hands before Sun God who holds saw and wears flounced garment. Between them: inscription. 19 mm.
e. Enthroned god turned to left holding ring and staff; beside him, suppliant goddess. Between them: two crossed bulls?, box of inscription below. 20 mm.
f. Front of suppliant goddess beside worshipper wearing long flounced garment, offering kid to goddess Ishtar who wears closed flounced garment; beside her only the hand of suppliant goddess preserved. In sky: head of nude bearded hero with curls; sun disc in crescent. 24 mm.
g. Three columns of inscription; figure with mace facing suppliant goddess (only hands visible). In sky: sun disc in crescent. 20 mm.
h. Goddess Ishtar holding scimitar in her lowered hand and double lion-headed club-mace and rein of her lion in other. Nude bearded hero holding flowing vase. In sky: human-headed fish (head missing), fish below. 21 mm.
CT 47, No. 67a.
Envelope; Samsuiluna, 4th year.

228.
a. Bow-legged dwarf, warrior brandishing weapons; lion attacking human victim. 16 mm.
b. Worshipper with clasped hands, beside second worshipper offering kid to Sun God. To right, conquering god brandishing weapons and stepping on dragon (only back visible). 15 mm. (Pl. XXXIV-7)
Louvre II, A.537; *ThD* No. 111.
Envelope; Samsuiluna, 4th year.

229.
Contest scene. Back of lion; nude bearded hero with curls triumphantly raising lion by hindquarters; human victim. 17 mm.
CT 47, No. 57a; *Harris*, p. 346.
Envelope; slave hire contract; Samsuiluna, 6th year.

230.
Figure with mace (head missing), facing suppliant goddess; between them: small figure with flowing vase (not clear), water rendered by drill holes. Worshipper led by interceding goddess toward god in ascending posture. 20 mm. (Pl. XXXIV-2)
Louvre II, A.543, Pl. 115, Fig. 3; *ThD* No. 117.
Envelope; Samsuiluna, 6th year.

231.
a. Front of figure with mace, conquering "king," stepping on enemy and brandishing multiple stylized club. Worshipper offering kid to Sun God (of latter, only front and hand visible). In sky: lion, sun disc in crescent. 15 mm.
b. God turned to left. Suppliant goddess turned to right. 13 mm.
c. Goddess full face; back of god turned to right. 20 mm.
Louvre II, A.544, Pl. 116, Fig. 2; *ThD* No. 120.
Envelope; Samsuiluna, 6th year.

232.
a. Part of god; figure with mace (head missing), facing suppliant goddess; back of figure. In field: lion, goat. 16 mm.
b. Front of worshipper; beside goddess facing deity (Ishtar), holding double lion-headed club (only serrated handle is visible). In field: double lion-headed club mace?, lion. Terminal: two columns of inscription. 20 mm.
Louvre II, A.549, Pl. 116, Fig. 3; *ThD* No. 22.
Tablet; Samsuiluna, 8th year.

233.
Figure with mace (head missing), facing suppliant goddess; between them: bow-legged dwarf. Two nude figures, one above the other, divided by horizontal line. Worshipper saluting enthroned god holding staff. Sun disc in crescent in sky. Terminal: mace? on long staff ending in circle. 20 mm.

Louvre II, A. 555, Pl. 116, Fig. 6; *ThD* No. 126.
Tablet; Samsuiluna, 8th year.

234.
Figure with mace (head missing), beside suppliant goddess who stands alongside bull-man offering kid to enthroned god who holds staff and sits on panelled throne placed on platform full-face goddess behind throne. 18 mm. (Pl. XXVIII-1)
Louvre II, A.552, Pl. 116, Fig. 8; *ThD* No. 127.
Tablet; Samsuiluna, 8th year.

235.
a. Lower part of deity's garment with grooved central wedge; double lion-headed club, staff composed of drillings; three columns for inscription; god with mace. 17 mm. (Pl. XXXIV-6)
b. Three columns of inscription; nude bearded hero with curls holding (vase?); water deity holds vase with water flowing in two streams into vases on ground. In sky: sun disc with drill holes in crescent. 22 mm. (Pl. XXVI-5)
CT 47, No. 58a; *Harris*, p. 63, 380.
Envelope; Samsuiluna, 9th year.

236.
a. God with mace facing suppliant goddess; between them sun disc in crescent; large eight-pointed star above two smaller ones. In lower half of seal, cluster of drill holes. 27 mm. (Pl. XXXIV-4)
b. God in ascending posture, behind him god with clasped hands, wearing single-horned crown and plain garment, with fringe indicated by drill holes as is also god's beard. Nude bearded hero with curls holding vase to chest; god holding staff, wearing single-horned crown topped with crescent, and flounced garment, stands on two reclining bulls? (heads missing). Between figures: small kneeling worshipper above box for inscription. Worshipper holding club across shoulder wears open long mantle. 24 mm. (Pl. XXIV-1)
c. Three columns of inscription; god with mace (missing), wearing long "swinging" beard. Arms and legs executed by drill holes. 24 mm. (Pl. XXXIV-5)
d. Figure with mace, facing suppliant goddess. 22 mm.
e. Worshipper offering kid to god in ascending posture brandishing curved weapon in left hand; suppliant goddess turned to right. In field: lion-fish. 19 mm.
f. Contest scenes: lion-griffin attacking human victim kneeling on double-tiered platform; nude bearded hero with curls kneeling on double-tiered platform, raises lion over his head; second hero, in profile, grasping bull in inverted position, only bull's head is visible. 20 mm. (Pl. XXXIII-8)
g. Nude bearded hero with curls extending vase with water flowing in two streams into two vases on ground, human-headed fish by hero's head. Deity with clasped hands beside worshipper offering kid to god enthroned on platform, holding staff. Beside throne: full-face goddess. In sky: sun disc in crescent; severed human head. 25 mm.
CT 47, No. 62; *Harris*, pp. 214, 278, 310.
Envelope; purchase of field; Samsuiluna, 9th year.

237.
a. Front of suppliant goddess behind worshipper offering kid to god enthroned on platform and holding staff. In field: severed human head, lion-fish below. 23 mm. (Pl. XXIX-5)
b. Possibly the terminal section of the seal. Two registers: above, two bull-men in contest; divided by horizontal line from seated goat with head turned back and human victim attacked by lion-griffin, below. In sky: sun disc in crescent. 23 mm.
c. Full-face goddess with clasped hands, beside full-face goddess with side locks. Worshipper offering kid to standing god who holds ring. In sky: two crossed animals (bulls), head of bull-man, sun disc in crescent, cow suckling calf; box of inscription below. 20 mm. (Pl. XXVIII-3)
d. Nude bearded hero with curls offering kid to Water Goddess with flowing vase. In sky: sun disc in crescent. Inscription scattered in field. 19 mm. (Pl. XXVI-8)
e. Suppliant goddess turned to left; two columns of inscription; worshipper carrying offering, turned to right. 20 mm.
Louvre II, A.553, Pl. 116, Fig. 7; *ThD* No. 131.
Envelope; Samuiluna, 10th year.

238.
a. Back of suppliant goddess, two columns of inscription. 23 mm.
b. Only one column of inscription visible; possibly a stylized lion-headed club. 18 mm.
c. Suppliant goddess (head obliterated), wearing flounced garment with grooved ridge; facing two columns for inscription. 15 mm. (Pl. XXXVI-1)
Louvre II, A.557; *ThD* No. 132.
Envelope; Samsuiluna, 10th year.

239.
a. Unidentifiable object beside Weater God holding lightning fork. 14 mm. (Pl. XXV-2)
b. Figure with mace (head missing). 14 mm.
Louvre II, A.555, Pl. 116, Fig. 10.
Tablet; Samsuiluna, 11th year.

240.
a. & b. Goddess Ishtar striding to left. Worshipper wearing flounced garment, carrying offering and turned to right. Scattered cuneiform signs in field. 25 mm. (Pl. XXV-6)
c. & d. Front of suppliant goddess, behind worshipper offering kid to Sun God (saw missing). Worshipper saluting standing god. In sky: sun disc in crescent. Scattered inscription in field. 21 mm. (Pl. XXXII-2)
e & f. Possibly the impression belongs to two seals but it is difficult to distinguish. Use of drill is prominent. God in ascending posture turned to left; two columns of inscription; suppliant goddess facing standing god who raises one hand; bull-man fighting lion. Cuneiform signs by legs of lion. 25 mm. (Pl. XXXI-1)
g. Lower part of garment and feet of deity, worshipper wearing long garment, with fringe executed by drill holes, offering kid to enthroned deity holding staff. Scattered inscription in field. 24 mm. (Pl. XXIX-6)
CT 47, 63a.
Envelope; bequest document; Samsuiluna, 14th year.

241.
a. Suppliant goddess turned to left (head and hands missing), with her back to Weather God holding lightning fork and standing on humped bull. Front part of figure (deity?), holding staff? Breakage in original seal. Ball-staff in field. 22 mm. (Pl. XXXV-3)
b. Three columns of inscription, possibly god with mace; spear behind his head. 17 mm.
Louvre II, A.556, Pl. 116, Fig. 9; *ThD* No. 136.
Envelope; Samsuiluna, 14th year.

242.
Possibly two impressions. Front of figure with mace, facing suppliant goddess, between them kneeling goat, lion club behind goddess. Second impression: small male figure behind suppliant goddess, beside bull-man offering kid to Sun God (saw missing). In sky: severed human head, sun disc in crescent. 15 mm. (Pl. XXVIII-9)
Louvre II, A.557, Pl. 117, Fig. 1; *ThD* No. 137.
Envelope; Samsuiluna, 18th year.

243.
a. Bull-man offering kid to enthroned god holding staff. In sky: sun disc in crescent. 15 mm. (Pl. XXVIII-10)
b. Weather God in ascending posture holding lightning fork. 16 mm. (Pl. XXV-4)
Louvre II, A.558; *ThD* No. 140.
Tablet; Samsuiluna, 22nd year.

244.
a. Figure with mace (head and mace obliterated), front of suppliant goddess: between them: two standing birds. 13 mm.
b. Weather God, brandishing weapon (missing), and holding lightning fork; facing another god. 17 mm. (Pl. XXXV-5)
c. Severed human head, front of suppliant goddess.
Louvre II, A.559; *ThD* No. 146.
Tablet; Samsuiluna, 33rd year.

245.
a. God Amurru, holding crook and stepping on gazelle? wearing "sun-hat" type crown and long garment composed of horizontal grooves (possibly a later addition). Suppliant goddess facing figure with mace. In field: small nude man wearing round cap, dividing line above kneeling worshipper in inverted position. Small nude female, star by her head; bow-legged dwarf below. 21 mm.
b. Back of goddess, turned to left, away from enthroned god holding ring and staff. Horns of deities' crowns indicated by drillings. In sky: two crossed bulls. 24 mm. (Pl. XXXV-6)
CT 45, No. 33.
Envelope; Samsuiluna.

246.
God turned to left in ascending posture, wearing flounced garment; figure with mace turned to right, wearing long "swinging" beard. 24 mm.
CT 47, No. 61a.
Envelope; purchase of slave; Samsuiluna.

247.
a. Nude priest with pail and sprinkler wearing round cap with upturned brim. 21 mm. (Pl. XXXI-10)
b. Two columns of inscription, worshipper offering kid to Sun God (only hand holding saw, and foot on artificial mountain are visible). 24 mm.
c. Old Assyrian seal. Hero wearing round cap kneeling on platform fighting with one of two crossed lions. 20 mm. (Pl. XLIV-9)
d. Goat sitting on knoll. 18 mm.
e. Worshipper offering kid to goddess Ishtar in profile; behind goddess, nude priest with pail and sprinkler standing on platform. In sky: vase, kneeling goat? In field: second kneeling goat with head turned back. 25 mm. (Pl. XXXI-8)
CT 47, No. 68a; *Harris*, pp. 63, 298, 373.
Envelope; legal document; Samsuiluna.

248.
a. Three columns of inscription; back of figure with mace? 18 mm.
b. Worshipper offering kid to Sun God. In sky: sun disc in crescent. 22 mm.

c. Back of human victim; small nude female; suppliant goddess beside worshipper (only back of latter visible), mongoose in field. 23 mm.
Louvre II, A.536.
Envelope; Samsuiluna.

249.
a. Worshipper offering kid to Sun God; between them, cow suckling calf. Terminal: nude bearded hero with curls triumphantly raising lion by hindquarters; below horizontal line, nude priest with pail and sprinkler (only the upper part visible). 18 mm. (Pl. XXXV-8)
b. Nude bearded hero with curls offering kid to god (only back of head and shoulders are visible). Between them: eight-pointed star. 12 mm. (Pl. XXXV-9)
Louvre II, A.560; *ThD* 1910, No. 147.
Tablet; Abi-eshuh.

250.
a. Four columns of inscription; also impression of gold cap decorated with granulation. 24 mm. (Pl. XXXVI-1)
b. Front of goddess with one hand raised wearing plain band round head. Four columns of inscription. 25 mm.
Louvre II, A.562; *ThD* 1910, No. 151.
Tablet; Ammiditana, 5th year.

251.
(Extensive use of drill in all impressions.)
a. Enthroned god on platform holding staff turned to left; in front of him sun disc in crescent, below small kneeling worshipper. Behind god, full-face long-haired goddess. Weather God brandishing curved weapon with one hand and holding lightning fork with other hand, places one foot on bull (head obliterated), and faces suppliant goddess. To left, bull-man offering kid (to enthroned god). Scattered inscription and star in field. 25 mm. (Pl. XXXVI-2)
b. Three columns of inscription; worshipper saluting enthroned deity who holds staff, wears single-horned crown and rests his feet on seated animal or monkey? 24 mm. (Pl. XXXVI-3)
c. Back of suppliant goddess turned to left; three columns of inscription; worshipper with one hand raised; sun disc of drill holes in crescent, miniature lion-headed demon brandishing weapon below. 20 mm. (Pl. XXXVI-4)
CT 45, 46; *Harris*, p. 42.
Tablet; litigation text; Ammiditana, 6th year.

252.
a. Worshipper offering kid to god in ascending posture. God holding double lion-headed club-mace, facing suppliant goddess. 18 mm.
b. Columns of inscription; back of figure with mace, drill technique apparent in arms. 15 mm.
c. Worshipper offering kid to full-face goddess standing on seated lions (Ishtar). 21 mm.
d. Four drill holes in sky; god with clasped hands turned to left. Nude priest with pail and sprinkler (missing), worshipper offering kid to god in ascending posture holding ring. Cuneiform sign in field. 18 mm. (Pl. XXXVII-3)
Louvre II, A.563; *ThD* No. 152.
Tablet; Ammiditana, 15th year.

253.
Impression in drilled style. Face and hand of worshipper facing god; second worshipper facing another god who holds staff with seven globes. Behind latter is Conquering God brandishing weapon; to right: back of a god. 15 mm. (Pl. XXXVII-1)
Louvre II, A.564, Pl. 117, Fig. 3; *ThD* No. 154.
Tablet; Ammiditana, 24th year.

254.
a. Drilled style impression; possibly nude bearded hero with curls holding staff with seven globes, another figure. Cuneiform signs. 18 mm. (Pl. XXXVII-2)
b. Two figures facing each other; between them: crescent, drill hole below. 15 mm.
c. Three columns of inscription; back of figure with mace. 15 mm.
Louvre II, A.565, Pl. 117, Fig. 5a; *ThD* No. 155.
Tablet; Ammiditana, 34th year.

255.
a. Worshipper facing god in ascending posture; behind latter figure of suppliant goddess without horned crown. Two nude bearded heroes with curls holding globe standard. 25 mm.
b. God wearing stylized horned crown topped by crescent. Conquering God brandishing mace. In field: ball-staff. 23 mm. (Pl. XXXVII-5)
Louvre II, A.566, Pl. 117, Fig. 6; *ThD* No. 156.
Tablet; Ammiditana, 37th year.

256.
a. Part of nude bearded hero with curls; worshipper offering kid to god in ascending posture; back of another god turned to right. 18 mm.
b. Suppliant goddess turned to left; worshipper turned to right, offering kid to deity holding double lion-headed club-mace (only the weapon is visible). 15 mm.
c. God with clasped hands. Figure with mace wearing

long "swinging" beard. 15 mm.
CT 45, No. 55; *Harris*, pp. 16, 47, 140.
Tablet: account of barley; Ammiditana.

257.
Three columns of inscription, Weather God brandishing weapon (missing). In his left hand he holds lightning fork and rein (made with drill holes) of bull which is his mount. 18 mm. (Pl. XXXVII-8)
Louvre II, A.256, Pl. 117, Fig. 7.
Tablet; Ammisaduqa, 8th year.

258.
Two columns of inscription. God in long plain garment with folded hands, wearing "sun-hat" type crown; goat-fish in field. 15 mm. (Pl. XXXVI-7)
Louvre II, A.569.
Tablet; Ammisaduqa, 8th year.

259.
a. Nude bearded hero with curls stands with clasped hands; worshipper wearing "sun-hat," facing figure holding mace; Conquering God holding weapon "symbol" made with numerous drill holes. Winged sphinx above lion. 18 mm. (Pl. XXXVII-10)
b. Head of god?, back of another?, enthroned deity, small kneeling worshipper; full-size worshipper. 15 mm. (Pl. XXXVI-6)
Louvre II, A.570, Pl. 117, Fig. 5; *ThD* No. 164.
Tablet; Ammisaduqa, 9th year.

260.
a. Back of small kneeling male figure. Worshipper facing Sun God in ascending posture. Weather God stepping on bull and holding fork with one hand and brandishing weapon with the other. 13 mm. (Pl. XXXVII-9)
b. Head of female worshipper. God with mace beside suppliant goddess; worshipper offering kid to Sun God (head missing) in ascending posture. In field: lightning fork, kneeling animal, human head; below: vase and ball-staff. Star disc in crescent. 21 mm.
Louvre II, A.571; Pl. 116, Fig. 11; Pl. 117, Fig. 8.
Tablet; Ammisaduqa, 16th year.

261.
a. Back of deity. Worshipper (head missing) wearing long garment, with fringes rendered by drill holes, faces full-face goddess. Between them: small figure of suppliant goddess. God beside bull-man. In field: cuneiform sign. 15 mm.
b. Drill holes (possibly belonging to standard). Goat standing on its hind legs. Head and arm of nude hero with curls; seated animal (goat) above. Extensive use of drill. 18 mm.
CT 45, No. 60; *Harris*, p. 67.
Tablet; legal document; Ammisaduqa, 17th year.

6.6. Larsa*

262.
Sun God holding saw, and placing one foot on artificial mountain rendered with incised lines. In sky. star disc in crescent. (A). 20 mm. (Pl. XXXIX-1)
Louvre II, A.479; *Jean* X, 2.
Tablet; sale of land; Sin-iqisham.

263.
a. Worshipper with one hand raised, wearing round cap with upturned brim. Folds of garment cut with deep incisions. 24 mm.
b. Sun God holding his saw vertically?, stepping on reclining bull (part obliterated). 26 mm.
Louvre II, A.480; *Jean* X, 3.
Tablet; division of estate; Sin-iqisham.

264.
(Sun) God turned to left placing one foot on artificial mountain. Figure with mace? turned to right. In sky: lightning fork. 22 mm. (Pl. XXXIX-3)
Louvre II, A.484; *Jean* X, 8.
Envelope; sale of land; Warad-Sin, 9th year

265.
Part of figure with mace. 27 mm.
Louvre II, A.481; *Jean* X, 9.
Envelope; sale of building; Warad-Sin, 9th year.

266.
a. Suppliant goddess beside worshipper carrying offering. Terminal: two columns of inscription. 22 mm.

*Letter in brackets denotes seal in *Louvre* II.

b. Suppliant goddess. 24 mm.
c. Part of fighting scene. Bull held in inverted position by nude hero. Part of lion. (D). 22 mm. (Pl. XXXIX-10)
d. Part of goddess Ishtar. Two columns of inscription. Suppliant goddess. 26 mm.
Louvre II, A.489; *Jean* X, 7.
Impression both on envelope and tablet; sale of house; Warad-Sin, 9th year.

267.
a. Sun God in ascending posture. Terminal in two registers: worshipper with one raised hand wearing round cap and long garment facing nude bearded hero with curls; below: two bull-men in contest. Seal was originally capped. (A). 24 mm. (Pl. XXXIX-2)
b. Nude priest on platform. Two columns of inscription. (C). 25 mm.
c. Figure with mace facing goddess Ishtar, rendered with careful detail, particularly the lion on which she steps. 24 mm. (Pl. XXXVIII-1)
Louvre II, A.490; *Jean* X, 8.
Envelope; sale of land; Warad-Sin, 9th year.

268.
a. Suppliant goddess beside worshipper offering kid to Sun God who places one foot on reclining human-headed bull. 20 mm.
b. Part of god. 14 mm.
c. Worshipper with clasped hands. He wears round cap and long simple garment. 18 mm.
Louvre II A.482; *Jean* X, 10.
Envelope and tablet; sale of building; Warad-Sin, 10th year.

269.
a. Nude hero in contest, his opponent is damaged. 26 mm.
b. Goddess Ishtar placing one foot on lion, holds only mace. 26 mm. (Pl. XXXVIII-6)
c. Goddess with face in profile holding mace. (A). 29 mm. (Pl. XXXVIII-9)
Louvre II, A.487; *Jean* X, 11.
Tablet; sale of building; Warad-Sin, 11th year.

270.
a. Goddess behind worshipper. (A). 23 mm.
b. Enthroned god on high-backed seat, holding staff? (B). 22 mm. (Pl. XXXIX-5)
Louvre II, A.483; *Jean* X, 12.
Tablet; sale of building; Warad-Sin, 12th year.

271.
a. Part of goddess Ishtar. Two columns for inscription. Weather God standing on lion-dragon (head obliterated). 25 mm. (Pl. XXXIX-7)
b. Figure with mace facing suppliant goddess. Terminal: two columns for inscription. 25 mm. (Pl. XL-1)
c. Female worshipper facing goddess Ishtar, box of inscription between them. Winged lion-headed demon to right of goddess. 25 mm. (Pl. XXXVIII-2)
Louvre II, A.485, Pl. 112, Fig. 10; *Jean* X, 15.
Tablet and envelope; sale of land; Warad-Sin.

272.
a. Suppliant goddess facing enthroned deity (large part is missing), who rests feet on reclining human-faced bull. In sky: disc in crescent. (C). 22 mm. (Pl. XXXIX-4)
b. Sun God placing one foot on artificial mountain. Behind god stands nude bearded hero with curls. (D). 23 mm.
c. Worshipper wearing round cap with upturned brim and long garment which opens at front; offering kid to goddess Ishtar who places one foot on lion. In sky: disc in crescent. 23 mm.
d. Figure with mace facing goddess Ishtar, of whom only double lion-headed club-mace and foot placed over well-engraved lion are visible. 26 mm. (Pl. XL-6)
e. Part of nude priest with pail and sprinkler. 16 mm.
f. Figure with mace. 22 mm.
Louvre II, A.486, Pl. 113, Fig. 1; *Jean* X, 14.
Envelope and tablet; sale of building; Warad-Sin.

273.
Part of goddess Ishtar with two columns of inscription. Figure with mace standing on platform engraved with mountain scales. 25 mm. (Pl. XL-5)
Louvre II, A.493; *Jean* X, 18.
Tablet; sale of building; Rim-Sin, 2nd year.

274.
Suppliant goddess behind worshipper, offering kid, possibly to goddess Ishtar, of whom only hand with mace is visible. 20 mm. (Pl. XXXVIII-7)
Louvre II, A.495; *Jean* X, 22.
Tablet; sale of land; Rim-Sin, 2nd year.

275.
Nude priest with pail and sprinkler. 21 mm.
Louvre II, A.496; *Jean* X, 28.
Tablet; account of grain; Rim-Sin, 2nd year.

276.
a. Goddess (Ishtar), wearing flounced garment and not carrying any weapons, stands over two reclining lions. Bull-man. Terminal: box for inscription above lion-fish. 23 mm. (Pl. XXXVIII-3)
b. Part of goddess Ishtar; part of second figure.

12 mm.
c. Suppliant goddess. 11 mm.
Louvre II, A.497; *Jean* X, 30.
Envelope; division agreement; Rim-Sin, 8th year.

277.
a. Goddess (Ishtar), standing on two lions. God with mace. In field: lion, bull above nude female; sun disc above cuneiform inscription. 22 mm. (Pl. XXXVIII-4)
b. Worshipper wearing long flounced garment, offering kid to goddess (Ishtar), also wearing flounced garment. Part of bull-man to right. In field: reclinining bull? above lion-fish.
Louvre II, A.598, Pl. 113, Fig. 2; *Jean* X, 29.
Envelope; sale of estate; Rim-Sin, 8th year.

278.
a. Figure with mace facing two columns of inscription; other side, suppliant goddess. 21 mm. (Pl. XL-4)
b. Full-faced goddess wearing square topped crown with one hand extended forward. Suppliant goddess. Both goddesses wear flounced garments with undulating lines. 22 mm. (Pl. XL-7)
Louvre II, A.515; *Jean* X, 33.
Tablet; sale of orchard; Rim-Sin, 9th year.

279.
a. Traces of god Shamash. 17 mm.
b. Traces of contest scene. 08 mm.
Louvre II, A.500; *Jean* X, 35.
Envelope; sale of palm tree orchard; Rim-Sin, 10th year.

280.
Figure with mace facing suppliant goddess; between them: monstrous mask. 26 mm. (Pl. XL-2)
Louvre II, A.501; *Jean* X, 36.
Envelope and tablet; sale of building; Rim-Sin, 10th year.

281.
a. Worshipper offering kid to Weather God who holds lightning fork and places one foot on reclining bull. (A). 24 mm. (Pl. XXIX-8)
b. Figure with mace facing suppliant goddess. (B). 24 mm. (Pl. XL-3)
Louvre II, A.503; *Jean* X, 41.
Envelope; sale of land; Rim-Sin, 15th year.

282.
a. Two columns of inscription, original seal had metal caps. 23 mm.
b. Part of god with mace. 17 mm.
Louvre II, A.506; *Jean* X, 46.
Envelope; sale of orchard; Rim-Sin, 17th year.

283.
a. Suppliant goddess beside worshipper offering kid to Ishtar? Only goddess' hand holding mace and foot placed on lion are visible. 24 mm. (Pl. XXXVIII-8)
b. Figure with mace. 21 mm.
Louvre II, A.507; *Jean* X, 48.
Envelope; payment, rental of orchard; Rim-Sin, 18th year.

284.
a. Two columns of inscription. Worshipper raises one hand in front of his face while offering kid with other. (B). 20 mm.
b. Damaged impression. Lion-headed demon, holding human victim. Goat above dividing line over small figure of an archer? God, wearing short mantle, holds down scimitar. (A). 21 mm. (Pl. XXXIX-9)
Louvre II, A.508, Pl. 114, Fig. 5; *Jean* X, 48.
Envelope; payment, rental of orchard; Rim-Sin, 20th year.

285.
a. Part of full-faced goddess. (B). 16 mm.
b. Part of Old Assyrian impression. Above, goat behind row of two worshippers. In field: vase above ball-staff. Below, row of reclining animals. (A). 16 mm. (Pl. XL-8)
Louvre II, A.509, Pl. 113, Fig. 3; *Jean* X, 67.
Envelope and tablet; sale of house; Rim-Sin, 25th year.

286.
Full-faced goddess (Ishtar) standing on two reclining lions. Worshipper. In sky: disc in crescent. Scattered cuneiform signs. 25 mm.
Louvre II, A.511; *Jean* X, 89.
Tablet; sale of orchard; Rim-Sin, 36th year.

287.
a. Figure with mace facing three columns for inscription. On other side: suppliant goddess. 21 mm. (Pl. XLI-2)
b. Figure with mace; behind him: three columns for inscription. In sky: disc in crescent. 22 mm. (Pl. XLI-1)
Louvre II, A.532, Pl. 115, Fig. 2; *Jean* XI, 192.
Envelope; receipt of date; Samsuiluna, 1st year.

288.
Three columns of inscription, figure with mace.

22 mm. (Pl. XLI-7)
Louvre II, A.535; *Jean* XI, 195.
Tablet; deposit on wool; Samsuiluna, 3rd year.

289.
Capped seal. Impression shows columns for inscription with part of figure with mace rendered in the drilled style. 20 mm. (Pl. XLI-6)
Louvre II, A.542; *Jean* XI, 211.
Envelope; loan of grain; Samsuiluna, 6th year.

290.
Figure with mace facing a serrated standard, possibly stylized double lion-headed club. Columns of inscription. 15 mm. (Pl. XLI-4)
Louvre II, A.546; *Jean* XI, 218.
Tablet; division of estate; Samsuiluna, 7th year.

291.
Two columns of inscription. Part of figure with mace. 25 mm. (Pl. XLI-8)
Louvre II, A.547; *Jean* XI, 217.
Tablet; recognition of debt; Samsuiluna, 7th year.

292.
Three columns of inscription. Two small male figures placed head to head (one upside down). 27 mm. (Pl. XLI-5)
Louvre II, A.548; *Jean* XI.
Tablet; loan of grain; Samsuiluna, 7th year.

293.
Three columns of inscription; figure with mace. 25 mm. (Pl. XLI-9)
Louvre II, A.533; *Jean* XI, 193.
Envelope; payment for palace produce; Samsuiluna.

294.
a. Trace of inscribed column; part of figure with mace. (A). 25 mm. (Pl. XLI-10)
b. Three columns of inscription. Suppliant goddess. (C). 25 mm. (Pl. XLI-3)
Louvre II, A.534; *Jean* XI, 196.
Tablet; loan of grain; Samsuiluna.

6.7. Ur.

295.
Only suppliant goddess visible. 22 mm. (Pl. XLIII-6)
IM 57408; *UET* 5, 478.
Tablet; allocation of rations; Gungunum, 19th year.

296.
Part of an incription. Worshipper, wearing long plain garment and round cap, raising right hand while left hand is hidden by garment. Bow-legged manikin in front, and vase over ball-staff behind him. 21 mm.
IM 57264; *UET* 5, 184.
Tablet and envelope; sale of slave girl; Sumu-ilum, 3rd year.

297.
Only terminal part of impression remains. Two registers. above, kneeling male figure behind attacking lion; below: lion attacking goat, back of second lion. 26 mm. (Pl. XLII-7)
IM 57225; *UET* 5, 103.
Tablet and envelope; partition of land; Sumu-ilum, 7th year.

298.
Coarse surface resulting in faint impression. Nude god with clasped hands and bent knees as if preparing to kneel, behind worshipper offering kid to enthroned "king." Disc in sky. Two columns of faint inscription. 20 mm. (Pl. XLIII-1)
IM 57611. *UET* 5, 171.
Tablet and envelope; sale of an orchard; Sumu-ilum, 9th year.

299.
Part of impression repeated all over tablet and damaged by inscription of tablet. Mythological scene of battle among gods: one god strangling second god with rope?, loser bent down, without crown, hands tied behind back. 17 mm. (Pl. XLIII-2)
IM 57318; *UET* 5, 299.
Tablet; loan contract; Sin-iqisham, 5th year.

300.
Bearded figure with mace. 25 mm.
IM 57375; *UET* 5, 414.
Tablet; loan of silver; Warad-Sin, 3rd year.

301.
Bareheaded worshipper wearing long garment, having both hands clasped, facing enthroned deity (part of garment and dais are visible). In sky: disc in crescent. 25 mm.

IM 57382; *UET* 5, 429.
Envelope; payment of money; Warad-Sin, 3rd year.

302.
Worshipper, wearing round cap and long plain garment, raising right hand. Other hand grasped by interceding goddess (only arm preserved). 20 mm.
IM 57319; *UET* 5, 330.
Tablet; loan contract; Warad-Sin, 4th year.

303.
Partly damaged impression. Worshipper wearing round cap with double brim and long garment, facing Sun God who places one foot on platform. 25 mm. (Pl. XLII-2)
IM 57337; *UET* 5, 331.

304.
a. Worshipper, wearing round brimmed cap and long garment opening at front, offering kid to Sun God who holds saw and places one foot on platform. In sky: star disc in crescent. 22 mm. (Pl. XLII-3)
b. Worshipper, wearing round brimmed cap and short kilt with point of fold between legs, offers kid to War Goddess (Ishtar) in profile holding bow and standing on two crouching lions; one foot placed on each. 22 mm. (Pl. XLIII-3)
IM 57221; *UET* 5, 92.
Tablet; adoption; Warad-Sin, 11th year.

305.
Standing god. Crescent in sky. Mongoose in field. 20 mm.
IM 57320; *UET* 5, 301.
Tablet; loan contract; Warad-Sin, 11th year.

306.
Worshipper wearing long plain garment raising right hand, other hand covered by garment, facing enthroned "king" on fleece stool, wearing similar garment and holding cup. Suppliant goddess behind "king." Terminal: three columns of inscription. 28 mm.
IM 57206; *UET* 5, 35.
Envelope; letter; Rim-Sin.

307.
Sun God in ascending posture placing one foot on platform and holding saw. 20 mm.
IM 57233; *UET*; 5, 122.
Tablet; exchange of premises; Rim-Sin, 1st year.

308.
Only terminal figures are visible. Two registers. Upper: hero holding weapon in front of him (shield), and brandishing curved weapon over his head. Second hero in posture of mastering an animal. Lower register: indistinct motif, two crossed lions. 25 mm.
IM 57233; *UET* 5, 122.
Tablet; exchange of premises; Rim-Sin, 1st year.

309.
Part of suppliant goddess behind god with mace who faces Sun God holding saw and placing one foot on platform (only part of god is visible). Vase over ball-staff in field. 25 mm. (Pl. XLII-4)
IM 57379; *UET* 5, 422.
Tablet and envelope; business document; Rim-Sin, 2nd year.

310.
Faint impression. Three columns of inscription. Suppliant goddess wearing necklace with counterweight. 22 mm.
IM 57383; *UET* 5, 430.
Tablet; receipt; Rim-Sin, 4th year.

311.
Impression of figure with mace repeated several times. 25 mm.
IM 57307; *UET* 5, 271.
Tablet; bridal gift; Rim-Sin, 7th year.

312.
Faint impression. Upper part of two worshippers.
IM 57345; *UET* 5, 338.
Tablet; loan contract; Rim-Sin, 10th year.

313.
Rough surface. Bareheaded worshipper wearing long garment, with clasped hands. Disc in crescent in sky. 22 mm.
IM 57376; *UET* 5, 415.
Tablet; business document; Rim-Sin, 10th year.

314.
Three columns of faint inscription. Part of deity visible. 27 mm.
IM 57449; *UET* 5, 536.
Tablet and envelope; payment account; Rim-Sin, 11th year.

315.
Incomplete and faint impression. Part of god. Terminal in two registers. Above: two nude females; below: two male figures facing each other (one figure partly damaged). Part of suppliant goddess. 27 mm. (Pl. XLII-9)
IM 57358; *UET* 5, 375.
Tablet; loan contract; Rim-Sin, 15th year.

316.
God with mace facing Sun God in ascending posture. Three drill holes in sky. Ball-staff and human head in field. 21 mm. (Pl. XLII-5)
IM 57359; *UET* 5, 376.
Tablet and envelope; loan contract; Rim-Sin, 17th year.

317.
Worshipper wearing round cap and long plain garment, offering kid. In field: double lion-headed club, mongoose. 23 mm.
IM 57271; *UET* 5, 199.
Envelope; rental of house; Rim-Sin, 26th year.

318.
Bull-man. Box of inscription above goat-fish. Original seal was capped. 23 mm. (Pl. XLIII-4)
IM 57385; *UET* 5, 432.
Tablet; receipt; Rim-Sin, 26th year.

319.
Lower part of Sun God? placing one foot on platform, turned to left. Figure with mace turned to right. 21 mm. (Pl. XLII-1)
IM 57292; *UET* 5, 238.
Envelope; hire of slave; Rim-Sin, 29th year.

320.
a. Nude bearded hero with curls, possibly in contest, turned to left. Turned to right, worshipper wearing cap with double brim and long garment opening at front, raising right hand in supplication while offering kid with other hand. 22 mm. (Pl. XLII-6)
b. Probably nude bearded hero with curls holding flowing vase, with fish swimming upstream. Part of suppliant goddess. 24 mm.
c. Nude bearded hero with curls. 22 mm.
IM 57235; *UET* 5, 126.
Tablet; business document; Rim-Sin, 31st year.

321.
God with mace behind worshipper wearing round cap with double brim and long plain garment, offering kid. Inscription between them. 22 mm.
IM 57589; *UET* 5, 821.
Tablet; record concerning cows; Rim-Sin, 31st year.

322.
Nude bearded hero with curls holding flowing vase, water flowing on one side into round vase. Usmu, the two-faced god, holding vase and standing on goat-fish. In sky: three drill holes. In field: pig. 24 mm. (Pl. XLIII-5)
IM 57594; *UET* 5, 826.
Tablet; legal document; Rim-Sin, 33rd year.

323.
Bull-man holding standard of double lion-headed club behind suppliant goddess. Figure with mace facing Sun God in ascending posture, holding saw. In field: vase above ball-staff. Star disc in crescent above mongoose. 25 mm.
IM 57390; *UET* 5, 437.
Tablet; receipt. Rim-Sin, 34th year.

324.
Suppliant goddess beside figure with mace facing goddess Ishtar. Part of impression is faint. Two columns of inscription. 25 mm.
IM 57389; *UET* 5, 436.
Tablet and envelope; receipt; Rim-Sin, 34th year.

325.
Suppliant goddess behind worshipper (only lower part of garment is visible. 24 mm.
IM 57300; *UET* 5, 252.
Envelope and tablet; legal document; Rim-Sin, 35th year.

326.
a. God Amurru holding crook and placing one foot on goat faces suppliant goddess. Between them: star disc in crescent above double lion-headed club. Terminal: winged lion-headed demon holding staff? in reverse position, above second lion-headed demon. 20 mm. (Pl. XLIII-7)
b. Part of suppliant goddess facing star. Below: drill hole above double lion-headed club. 17 mm. (Pl. XLIII-8)
IM 57304; *UET* 5, 256.
Tablet; legal document; Samsuiluna, 23rd year.

7

BIBLIOGRAPHY

ABBADA, K. M.

1974 "Recent Archaeological Acquisitions." *Sumer* 30.

AHMED, Abdul-Karim Abdullah

1967 "The Paramount God and the Old Name of Al-Dhiba'i." *Sumer* 23.

AL-HASHIMI, Ridha

1972 "New Light on the Date of Harmal and Dhiba'i." *Sumer* 28.

ALSTER, Bendt

1971-72 "Ninurta and the Turtle, UET 6/12." *JCS* 24.

AMIET, Pierre

1960 "Notes sur le répertoire iconographique de Mari à l'époque du palais." *Syria* 37.

BAQIR, Taha

1950 "Another Important Mathematical Text." *Sumer* 6, no. 2.

1950 "Mathematical Text." *Sumer* 6, no. 1.

1949 "Supplement to the Date-Formulae from Harmal." *Sumer* 5, no. 2.

1949 "Date Formulae and Date-Lists." *Sumer* 5, no. 1.

1948 "A New Law-Code from Tell Harmal." *Sumer* 4.

1946 "Excavations at Tell Harmal." *Sumer* 2, no. 2.

BAQIR, Taha, and MUSTAFA, M. A.

1945 "Iraq Government Sounding at Der." *Sumer* 1, no. 2.

BARRELET, Marie-Thérèse

1968 *Figurines et reliefs en terre-cuite de la Mésopotamie antique, I.* Institut français de Beyrouth; Bibliothèque archéologique et historique. Paris, 1968.

1955 "Les déeses armées et ailées." *Syria* 32.

BIGGS, Robert D.

1969 Review of Figulla, H. H., *Cuneiform Texts from Babylonian Tablets in the British Museum: Part XLVII: The Old Babylonian Nadītu Records. JNES* 28.

BÖRKER-KLÄHN, Jutta

1975 "Sulgi badet." *ZA* 64.

BUCHANAN, Briggs

1971 "A Snake Goddess and her Companions." *Iraq* 33.

1970 "Cylinder Seal Impressions in the Yale Babylonian Collection Illustrating a Revolution in Art, circa 1700 B.C." *Yale Library Gazette* 45.

1960 "The Cylinder Seals of the Yale Babylonian Collection." *Yale Library Gazette* 35.

CIVIL, M.

1961 "The Home of the Fish." *Iraq* 23.

COLLON, Dominique

1975 *The Seal Impressions from Tell Atchana/Alalakh. AOAT* 27.

COOPER, Jerrold S.
1975 "Heilige Hochzeit." *RlA* 4.

DALES, George F.
1963 "Necklaces, Bands and Belts on Mesopotamian Figurines." *RA* 57.

DALEY, S., WALKER, C. and HAWKINS, D.
1976 *The Old Babylonian Tablets from Tell al-Rimah.* London.

DELOUGAZ, Pinhas
1952 *Pottery from the Diyala Region.* OIP LXIII, Chicago.

DURING-CASPERS, E. C. L.
1971-72 "The Gate-Post in Mesopotamian Art: a short Outline of its Origin and Development." *Jaarbericht, Ex Orient Lux* 22.

ELLIS, Maria de Jong
1974 "The Division of Property at Tell Harmal." *JCS* 26.
1969 "Taxation and Land Revenues in the Old Babylonian Period." Ph.D. Dissertation, Yale University.

FAUST, D. E.
1941 *Contracts from Larsa dated in the reign of Rim-Sin.* Yale Oriental Series, Babylonian Texts, VIII.

FIGULLA, H.
1914 *Altbabylonische Verträge.* Vorderasiatische Schriftdenkmäler der Königlichen Museen zu Berlin 13. Leipzig.

FINKELSTEIN, J. J.
1972 *Late Old Babylonian Documents and Letters.* Yale Oriental Series, Babylonian Texts 13. New Haven.

FRANKFORT, Henri
1943 *More Sculpture from the Diyala Region.* OIP LX, Chicago.

FRANKFORT, Henri, LLOYD, Seton, and JACOBSEN, Thorkild
1940 *The Gimilsin Temple and the Palace of the Rulers at Tell Asmar.* OIP XLIII, Chicago.

GADD, C. J.
"Hammurabi and the End of his Dynasty." *Cambridge Ancient History*, Vol. II, Pt. 1.

GIBSON, McGuire, and BIGGS, Robert D.
1977 *Seals and Sealing in the Ancient Near East.* Malibu.

GOETZE, Albrecht
1958 "Fifty Old Babylonian Letters from Harmal." *Sumer* 14.
1950 "Sin-iddinam of Larsa, new tablets from his reign." *JCS* 4.
1948 "The Laws of Ashnunna Discovered at Tell Harmal." *Sumer* 4, no. 2.

HALDAR, Alfred
1952 "On the Wall Painting from Court 106 of the Palace of Mari." *Orientalia Suecana* 1.

HALLO, William W., and SIMPSON, W. K.
1971 *The Ancient Near East.* New York.

HARRIS, Rivkah
1955 "The Archive of the Sin Temple in Khafajeh." *JCS* 9.

KARIM, A. A.
1967 "The Paramount God and the Old Name of Al-Dhiba'i." *Sumer* 23.

KINGSBURY, E. C.
1963 "A Seven-day Ritual in the Old Babylonian Cult at Larsa." *HUCA* 34.

KUPPER, J. K.
1961 *L'iconographie du dieu Amurru, dans la glyptique de la 1 re. dynastie babylonienne.* Mémoires de l'Académie Royale de Belgique, C1. des lettres 55, No. 1. Bruxelles.

LANDSBERGER, Benno
1948 *Sam 'al; Karatepe harabelerinin keşti ile ilqili araştirmaler*. Ankara.

LEEMANS, W. F.
1968 "Old Babylonian Letters and Economic History." *JESHO* 11.

LYON, D. G.
1906 "The Seal Impressions on an Old Babylonian Contract." *JAOS* 27.

MATOUŠOVA, M.
1966 "Running Adad." *Sumer* 22.

MAXWELL-HYSLOP, K. R.
1971 *Western Asiatic Jewellry c. 3000-612 B.C.* London.

MOOREY, P. R. S.
1975 "The Terracotta Plaques from Kish and Hursagkalama c. 1850 B.C." *Iraq* 37.

MOOREY, P. R. S., and GURNEY, O. R.
1973 "Ancient Near Eastern Seals at Charterhouse." *Iraq* 35.

MOORTGAT, Anton
1969 *The Art of Ancient Mesopotamia*. London.

MOORTGAT-CORRENS, Ursula
1968 "Die ehemalige Rollsiegel-Sammlung Erwin Oppenländer." *Baghdader Mitteilungen* 4.

MUSTAFA, M. A.
1948-49 "Sounding at Tell al-Dhiba'i." *Sumer* 5.

NAGEL, Wolfram
1957-58 "Glyptische Probleme der Larsa-Zeit." *AfO* 18.

OPIFICIUS, Ruth
1961 *Das altbabylonische Terrakottarelief.* Berlin.

OZGUC, Nimet
1968 *Seals and Seal Impressions of Level Ib, from Karum Kanish*. (Türk Tarih Kurumu Yayinlarindan 5, no. 25.) Ankara.
1965 *The Anatolian Group of Cylinder Seal Impressions from Kültepe.* (TTKY 5, no. 22.) Ankara.

PARKER, Barbara
1975 "Cylinder Seals from Tell al-Rimah." *Iraq* 37.

PARROT, André
1960 *The Arts of Mankind*. London.
1933 "Les fouilles de Tello et de Senkereh-Larsa (campagne 1932-1933)." *RA* 30.

PIET, J.
1969 "An Old Babylonian Crystal Seal." *Journal of the Ancient Near Eastern Society of Columbia University* 2.

PINCHES, T.
1964 *Old Babylonian Business Documents*. Cuneiform Texts from Babylonian Tablets in the British Museum, XLV. London.

PORADA, Edith
1979 "The Iconography of Death in Mesopotamia in the Early Second Millennium B.C." 26th Rencontre assyriologique internationale.
1975 "Karum." *RlA* 4.
1974/77 "Die Siegelzylinder-Abrollung auf der Amarna-Tafel BM 29841 im Britischen Museum." *AfO* 25.
1971 "Remarks of Seals found in the Gulf States." *Artibus Asiae* 33.
1966 "Les cylindres de la jarre Montet." *Syria* 43.

PORADA, (Continued)

1964 "Iconography and Style of a Cylinder Seal from Kantara in Cyprus." Moortgat festschrift. Kurt Bittel, ed., *Vörderasiatische Archäologie.* Berlin.

1964 "An Emaciated Male Figure of Bronze in the Cincinnati Art Museum." *Studies Presented to A. Leo Oppenheim* (Chicago).

1957 "Syrian Seal Impressions on Tablets Dated in the Time of Hammurabi and Samsu-iluna." *JNES* 16.

1947 *Seal Impressions of Nuzi.* AASOR 24 (1944-1945). New Haven.

PORADA, Edith, and BASMACHI, F.

1951 "Nergal in the Old Babylonian Period." *Sumer* 7, pt. 1.

RANKE, H.

1906 *Babylonian Legal and Business Documents from the Time of the First Dynasty of Babylon, chiefly from Sippar.* BE Series A: Cuneiform Documents. Vol. 6, pt. 1. Philadelphia.

READE, J. E.

1973 "Tell Taya (1972-73): Summary Report." *Iraq* 35.

SEYRIG, H.

1955 "Antiquites syriennes 60." *Syria* 32.

SIMMONS, S. D.

1958 "Early Old Babylonian Tablets from Harmal and Elsewhere." *JCS* 13.

SOLLBERGER, Edmond

1969 "Old Babylonian Worshipper Figurines." *Iraq* 31.

SOLLBERGER, Edmond, and KUPPER, J. G.

1971 *Inscriptions royales sumériennes et akkadiennes.* Paris.

SPYCKET, Agnes

1960 "La déesse Lama." *RA* 54.

1948 "Un élément de la parure féminine à la 1re dynastie babylonienne." *RA* 42.

THUREAU-DANGIN, F.

1918 "La Chronologie de la dynastie de Larsa." *RA* 15.

VAN BUREN, E. D.

1959 *Catalogue of the Ugo Sissa Collection.* Rome.

1952 "Homage to a Deified King." *ZA* 50.

1949 "The Rod and Ring." *Archiv Orientalni* 17.

1945 *Symbols of the Gods in Mesopotamian Art.* (Analecta Orientalia 23.) Rome.

1933 "The God Ningizzida." *Iraq* 1.

WATERMAN, L.

1912-13 "Business Documents of the Hammurapi Period." *AJSL* 29.

WEIHER, E. von

1971 "Der babylonische Gott Nergal." *AOAT* 11.

WISEMAN, D. J.

1960 "The Goddess Lama at Ur." *Iraq* 22.

ADDENDUM

Since the manuscript was submitted for publication a number of relevant articles and books have been published. They are listed below.

AL-JADIR, W., and L. AL-GAILANI
1981 "Seal Impressions from Sippar." *Sumer* 37, 129-44.

AL-JADIR, W., and L. AL-GAILANI
1987 "Seal Impressions from Sippar II." In *Researches on the Antiquities of Saddam Dam Basin Salvage and Other Researches.* Baghdad, 163-44.

COLLON, D.
1986 *Catalogue of the Western Asiatic Seals in the British Museum. Cylinder Seals III: Isin-Larsa and the Old Babylonian Periods.* London.

GREEN, A.
1986 "The Lion-Demon in the Art of Mesopotamia and Neighboring Regions." *Baghdader Mitteilungen* 17, 141-254.

KLENGEL-BRANDT, E.
1983 "Siegelabrallungen auf altbabylonischen Tontafeln aus Babylon." *Altorientalisch Forschungen* 10, 65-106.

PLATES

1 (1.a)

2 (4.c)

3 (3.b)

4 (27.c)

5 (66)

6 (115)

7 (4.d)

10 (3.a)

8 (25.a)

11 (61)

12 (49.a)

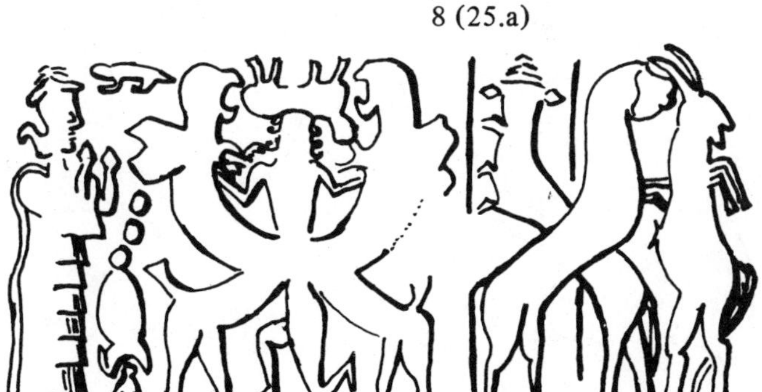
9 (19)

13 (69.m)

1 (6)

2 (42.a)

3 (42.b)

4 (4.b)

5 (10.a)

6 (16.e)

7 (33.b)

8 (41.a)

9 (63.b)

10 (69.a)

11 (37)

12 (30)

1 (57.a)

2 (60.a)

3 (60.b)

4 (60.c)

5 (3.c)

6 (31.a)

7 (2.a)

8 (23)

9 (69.c)

1 (1.b)

2 (16.f)

3 (59.a)

4 (76.a)

5 (35)

6 (61)

7 (69.b)

8 (85)

1 (9)

2 (27.b)

5 (11.a)

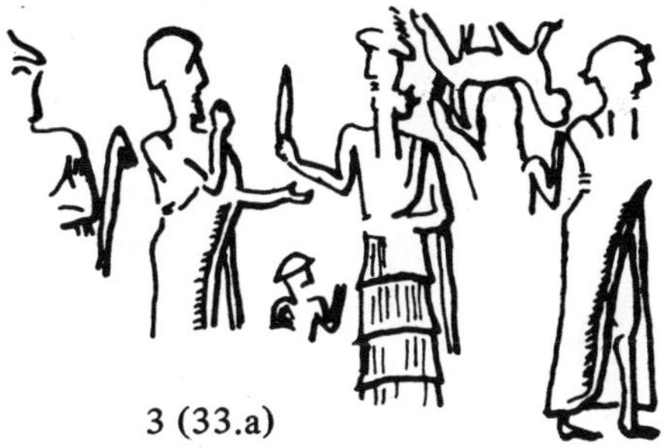

3 (33.a)

6 (2.b)

4 (69.f)

7 (5)

9 (134.a)

8 (38)

1 (11.b)

2 (12.b)

3 (20.d)

4 (46.c)

5 (62.d)

6 (108.b)

7 (139)

8 (53.a)

9 (54.c)

10 (137.b)

1 (12.a)

2 (20.c)

4 (137.a)

3 (45)

5 (10.c)

6 (27.a)

7 (20.a)

8 (135.b)

9 (147)

10 (152.b)

1 (34.b)

2 (150.b)

4 (111)

3 (90)

5 (108.a)

6 (48.c)

7 (55.a)

8 (14)

9 (104.a)

10 (39.c)

1 (8.b)

2 (25.c)

3 (50.a)

4 (51.a)

5 (54.a)

6 (54.b)

7 (56.a)

8 (62.b)

9 (64)

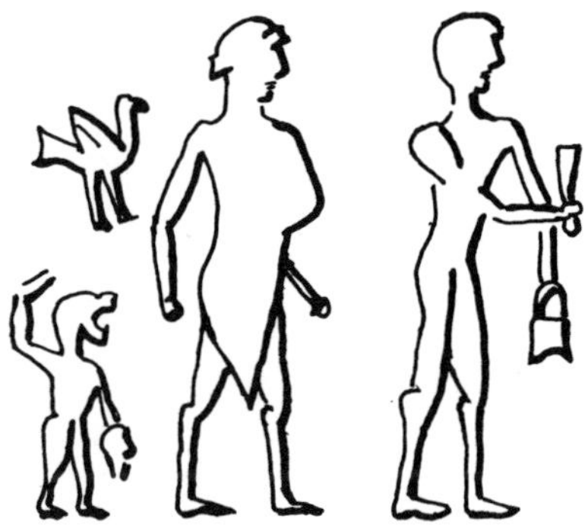

1 (20.b)

2 (34.a)

3 (48.b)

4 (50.c)

5 (53)

6 (56.b)

7 (62.a)

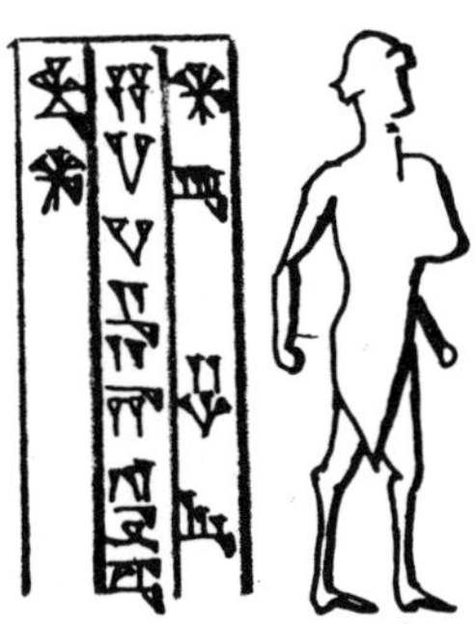

8 (54.f)

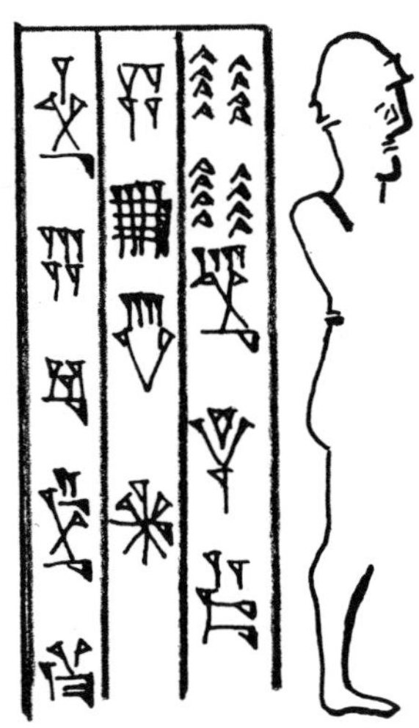

11 (75)

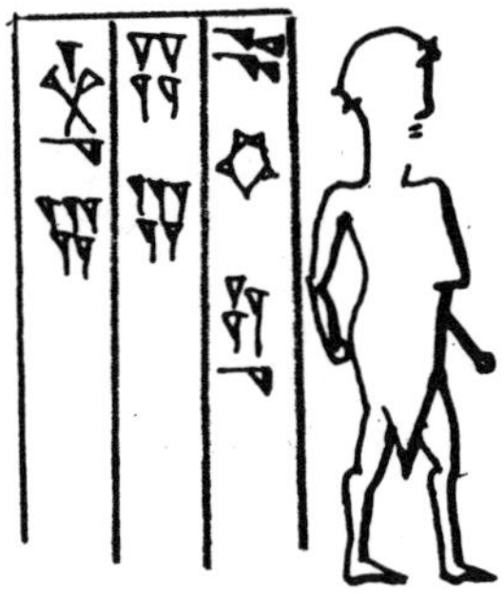

9 (73)

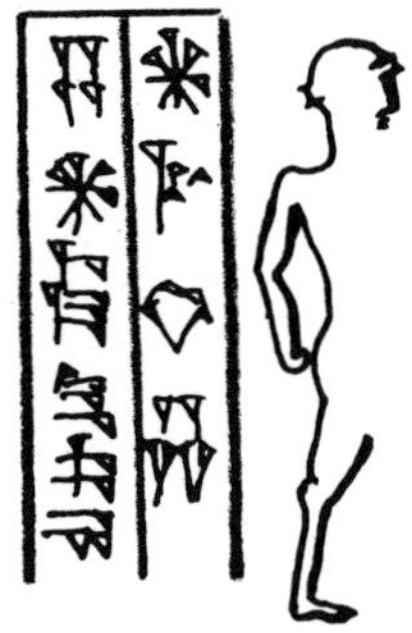

10 (74)

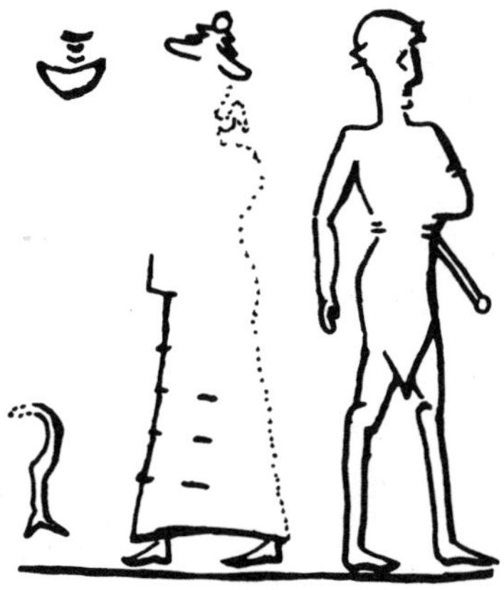
1 (77.a)

2 (78)

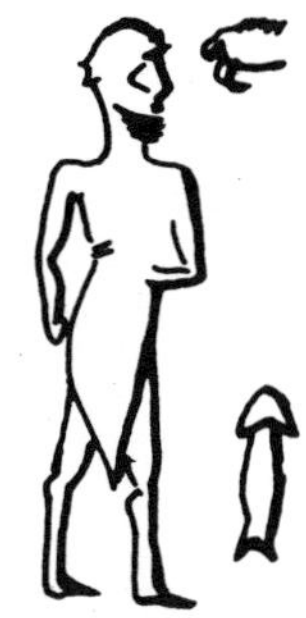
3 (80.a)

4 (91)

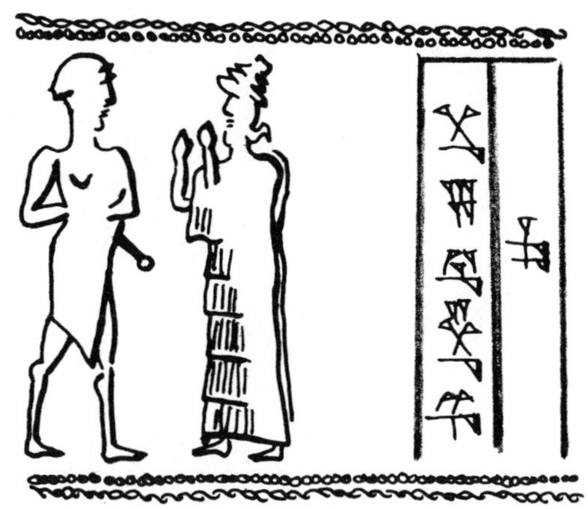
5 (79)

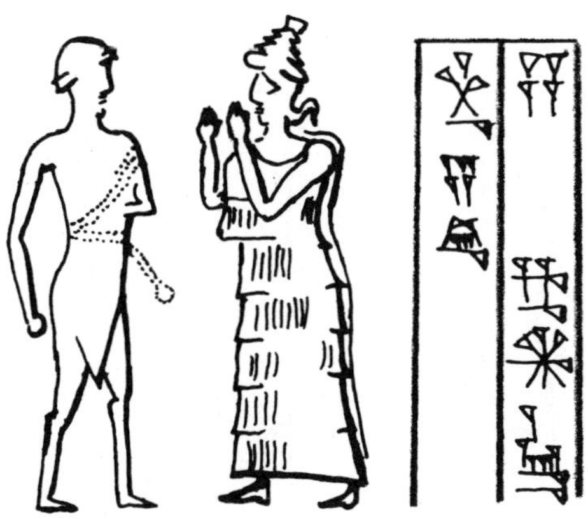
6 (88.a)

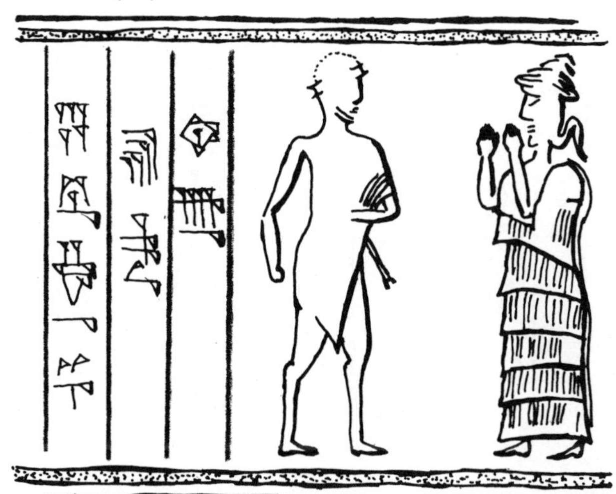
7 (89)

8 (94)

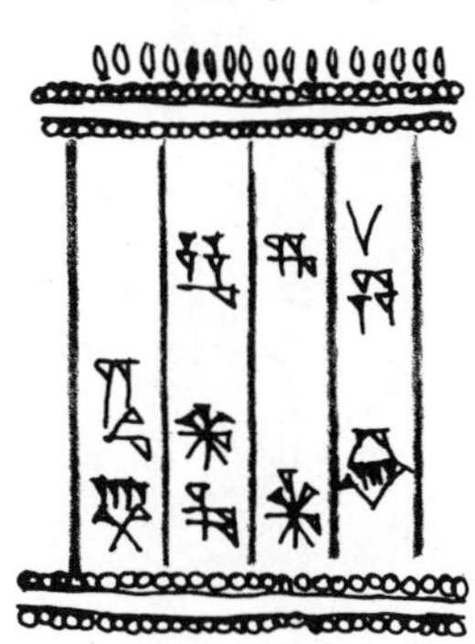
9 (83)

10 (88.b)

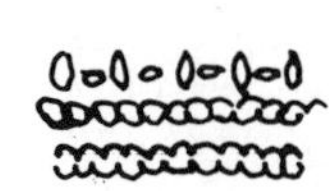
11 (93.b)

1 (114.c)

2 (114.a)

3 (113)

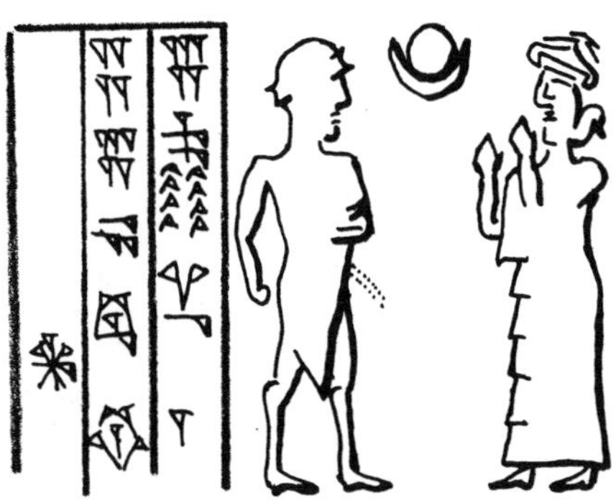

4 (99.a)

5 (95)

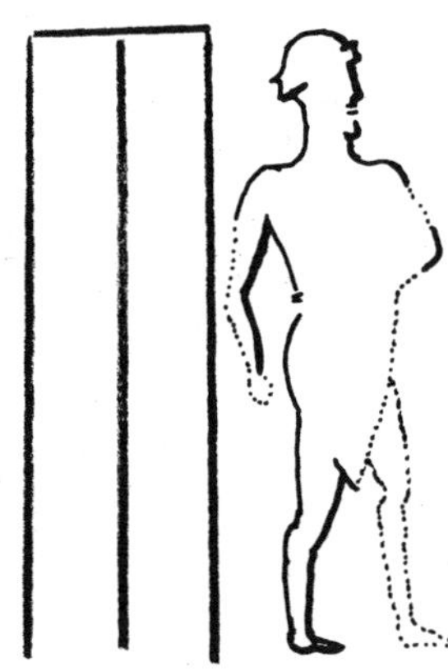

6 (98)

7 (103)

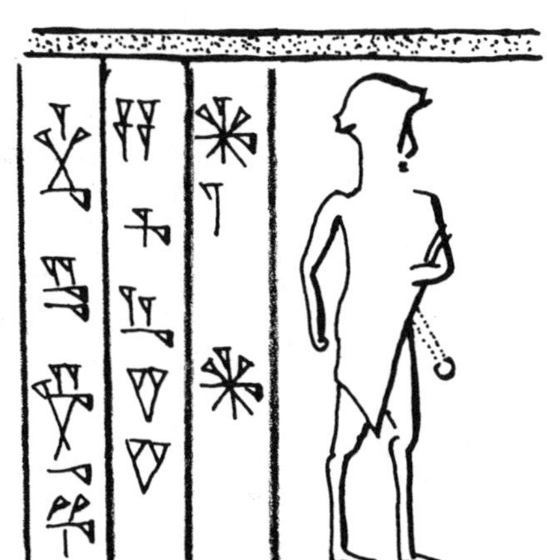

8 (107)

1 (57.c)

2 (65.b)

3 (51.b)

4 (47.a)

5 (69.n)

6 (18)

7 (26.a)

8 (100)

9 (106)

1 (16.a)

2 (17.a)

3 (13.a)

4 (138)

6 (57.b)

5 (48.a)

7 (47.b)

9 (109.b and a)

8 (43)

10 (101)

1 (123.a)

3 (128.b.1)

4 (128.b.2)

2 (129.h)

5 (130.m)

6 (130.f)

7 (131.n)

8 (128.g)

9 (128.h)

10 (128.0)

12 (130.i)

13 (129.i)

11 (128.p)

15 (130.j)

14 (131.r)

16 (131.s)

1 (126)

2 (125)

3 (131.h)

4 (131.d)

5 (131.g)

6 (131.f)

7 (131.a)

1 (130.k)

2 (131.j)

3 (131.i)

4 (129.c)

5 (130.b)

6 (119)

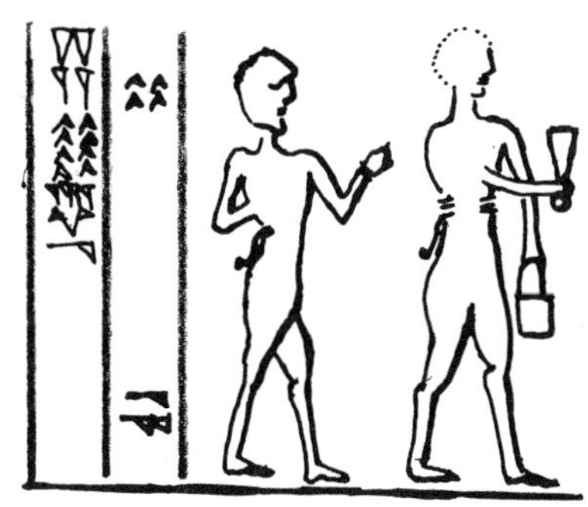
7 (117)

8 (129.e)

9 (39.a)

10 (121)

11 (131.t)

12 (122)

1 (128.j)

2 (129.a)

3 (128.l)

4 (128.m)

5 (131.c)

6 (129.d)

7 (129.f)

8 (123.b)

1 (116)

2 (118)

3 (129.g)

4 (131.b)

5 (130.a)

6 (130.d)

7 (129.n)

1 (128.d)

2 (131.k)

3 (131.m)

4 (131.o)

6 (129.k)

7 (130.g)

5 (120)

9 (128.c)

8 (130.e)

10 (128.e)

11 (130.c)

12 (124.b)

1 (4.a)

2 (36.a)

3 (39.a)

4 (24)

5 (44)

6 (135.a)

7 (135.d)

8 (140.b)

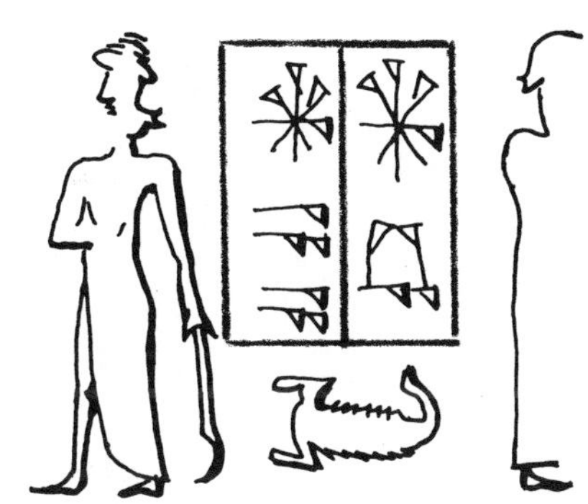
9 (144)

10 (150.a)

11 (146.b)

1 (182.b)

2 (182.c)

3 (182.d)

4 (182.e)

5 (183)

1 (185.b)

2 (157.b)

3 (163)

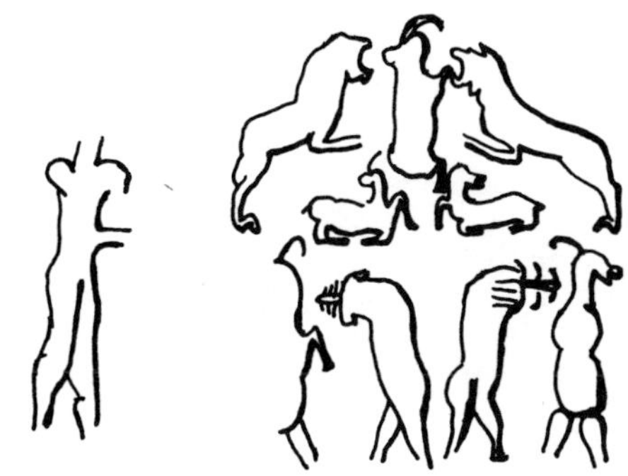

4 (216.c)

5 (168)

6 (185.c)

7 (158.a)

8 (159.a)

9 (159.d)

1 (167)

2 (164)

3 (166.a)

4 (156)

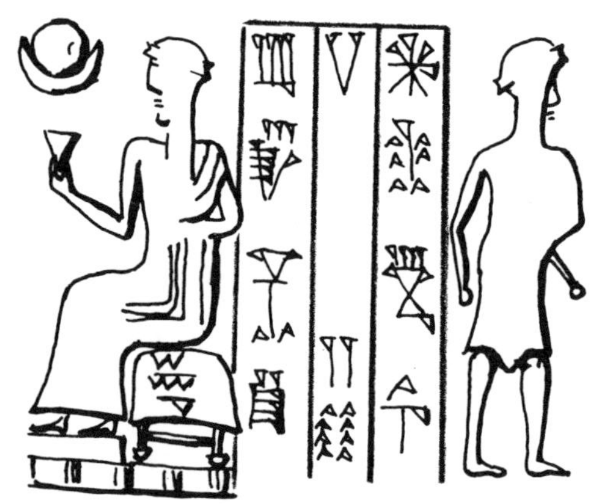

5 (182.a)

6 (184.a)

7 (170.c)

8 (186.a)

1 (201.h)

2 (189.d)

3 (206.f)

4 (201.d and e)

5 (188.e)

6 (240.a and b)

8 (201.a, b, and c)

7 (198.i)

9 (216.d)

1 (206.c)

2 (216.b)

3 (211.a)

4 (198.e)

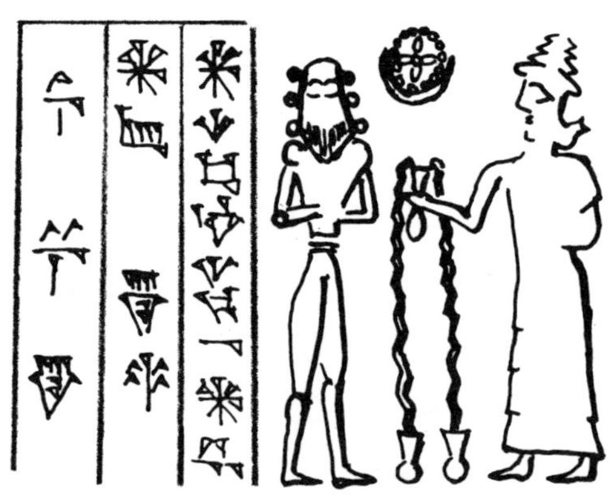
5 (235.b)

6 (212.d)

7 (198.f)

8 (237.d)

1 (190.i)

2 (202.b)

3 (209.a)

4 (208.b)

5 (190.e)

6 (190.g)

7 (210.b and c)

1 (234)

2 (221.e)

3 (237.c)

4 (226.b)

5 (216.b)

6 (212.e)

7 (224.c)

8 (216.i)

9 (242)

10 (243.a)

1 (217)

2 (206.a)

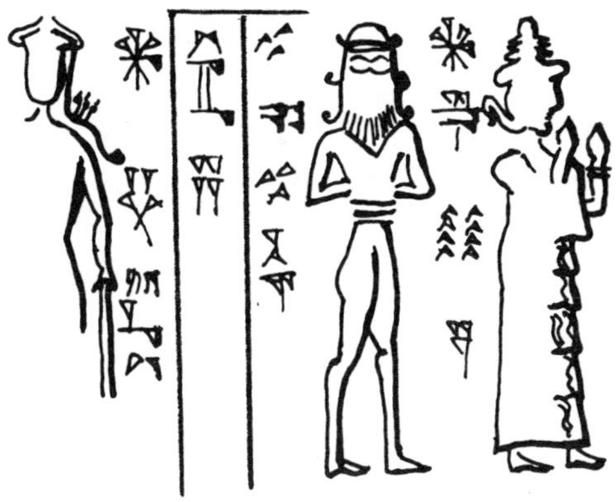

3 (195.h)

4 (218.a)

5 (237.a)

6 (240.g)

7 (199.a)

1 (187.f)

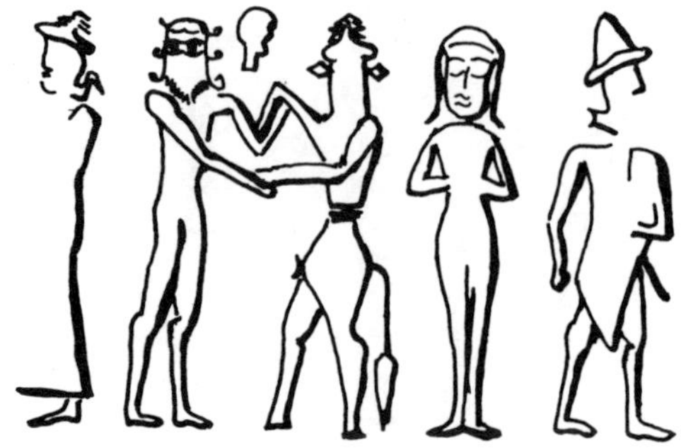

2 (195.a)

3 (197.a)

4 (198.b)

5 (201.g)

6 (203.4)

7 (206.f)

8 (211.d and e)

9 (208.d and e)

1 (240.e and f)

3 (220)

2 (216.e)

4 (222.f)

5 (200.a and b)

6 (196.d)

7 (207.a)

8 (247.e)

9 (221.d)

10 (247.a)

1 (195.b)

2 (240.c and d)

3 (205.b)

4 (208.c)

5 (194.a)

6 (197.c)

7 (216.k)

8 (216.j)

1 (195.g)

2 (154.c)

3 (206.b)

4 (189.b)

5 (188.c)

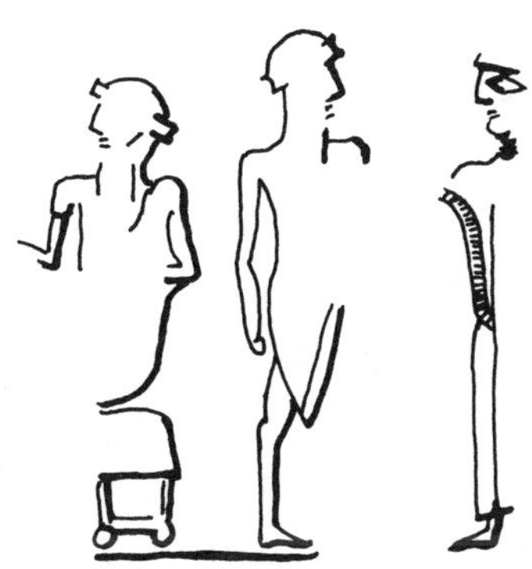
6 (222.d)

7 (202.c)

8 (236.f)

1 (236.b)

2 (230)

3 (226.c)

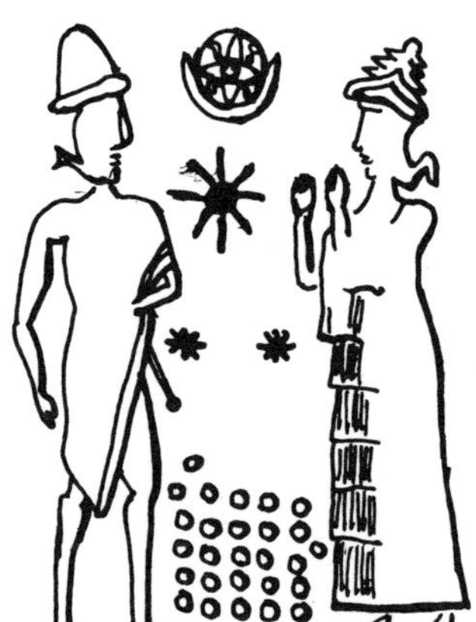

4 (236.a)

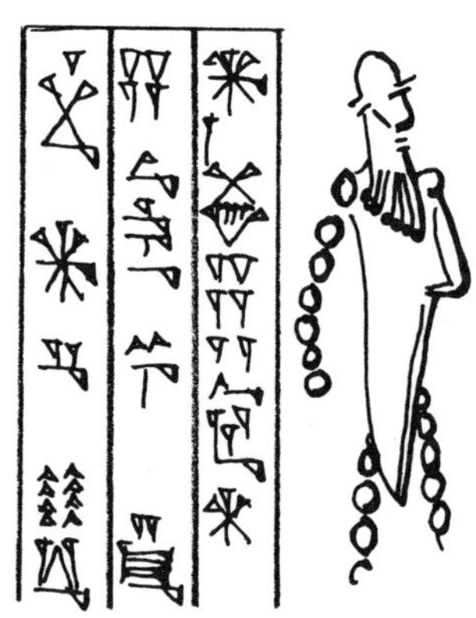

5 (236.c)

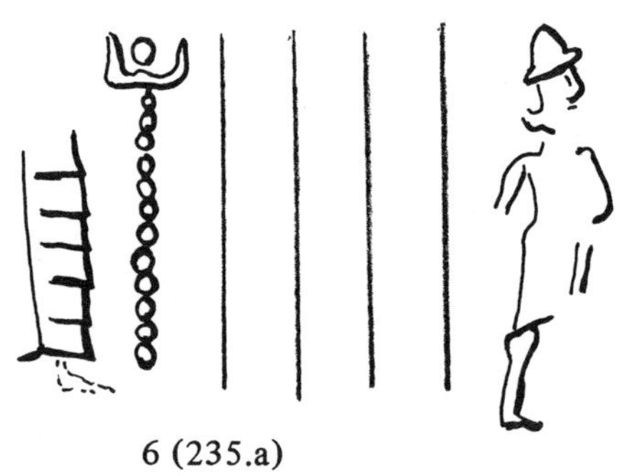

6 (235.a)

7 (228.b)

8 (224.a)

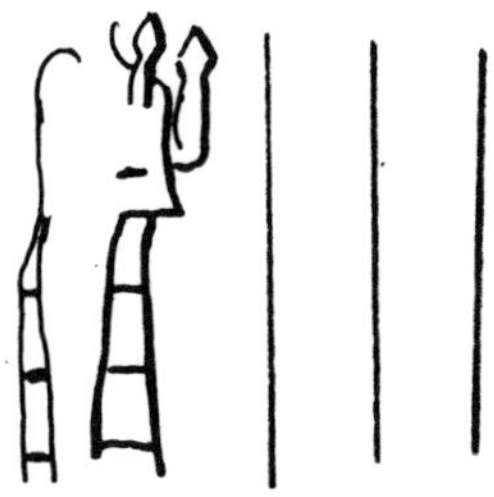

1 (238.c)

2 (239.a)

3 (241.a)

4 (243.b)

5 (244.b)

6 (245.b)

7 (245.a)

8 (249.a)

9 (249.b)

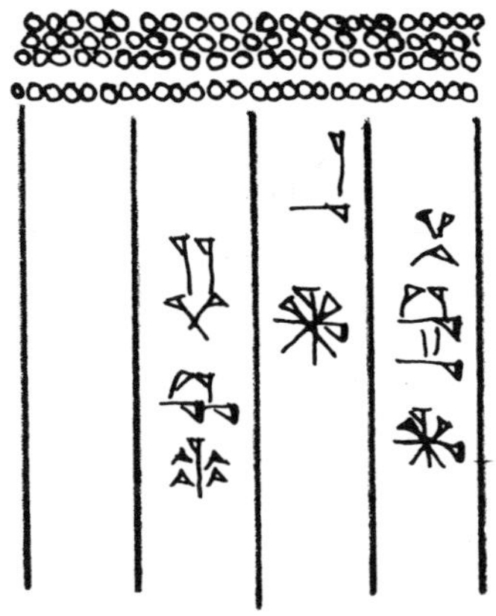

1 (250.a)

2 (251.a)

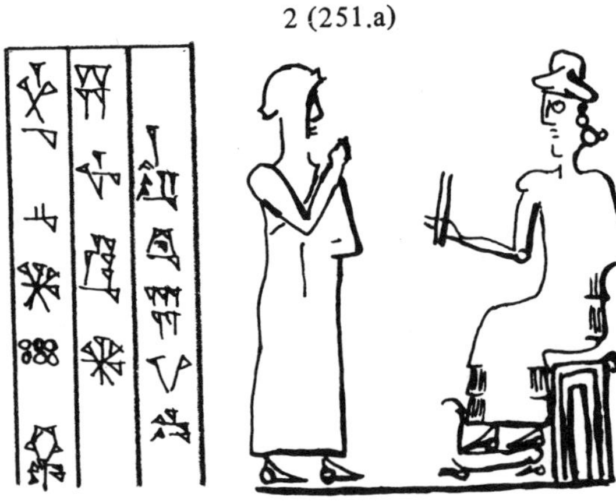

3 (251.b)

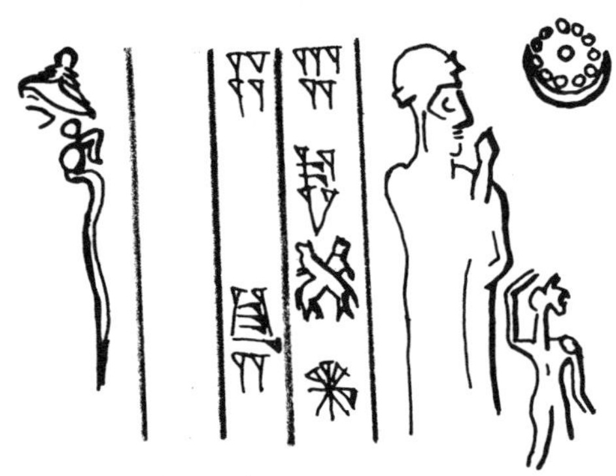

4 (251.c)

5 (172.b)

6 (259.b)

7 (258)

8 (179.a)

9 (179.b)

1 (253)

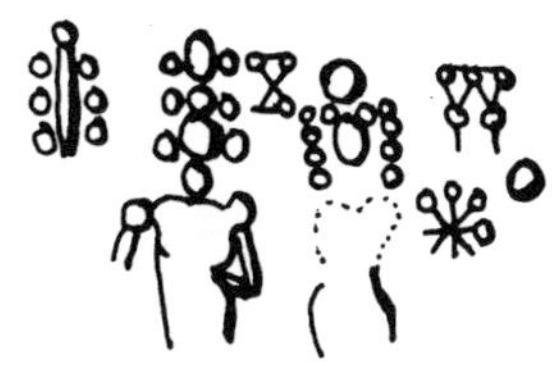

2 (254.a)

3 (252.d)

4 (173)

5 (255.b)

6 (179.c)

7 (180)

8 (257)

9 (260.a)

10 (259.a)

11 (209.c)

12 (175)

1 (267.c)

2 (271.c)

3 (276.a)

4 (277.a)

5 (277.b)

6 (269.b)

7 (274)

8 (283.a)

9 (269.c)

1 (262)

2 (267.a)

3 (264)

4 (272.a)

5 (270.b)

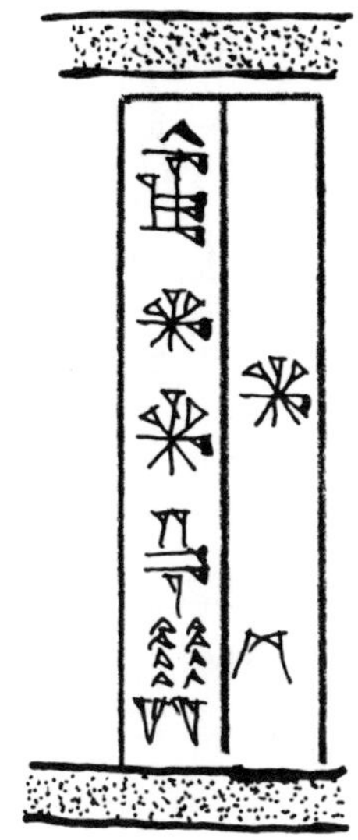
6 (282.a)

7 (271.a)

8 (281.a)

9 (284.b)

10 (266.c)

1 (271.b)

2 (280)

3 (281.b)

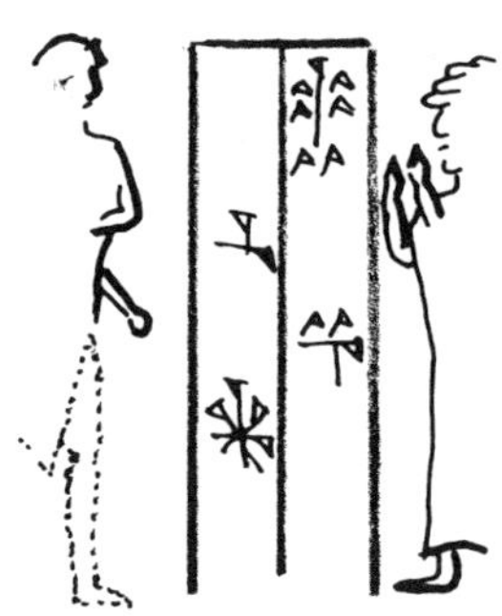

4 (278.a)

5 (273)

6 (272.d)

7 (278.b)

8 (285.b)

1 (287.b)

2 (287.a)

3 (294.b)

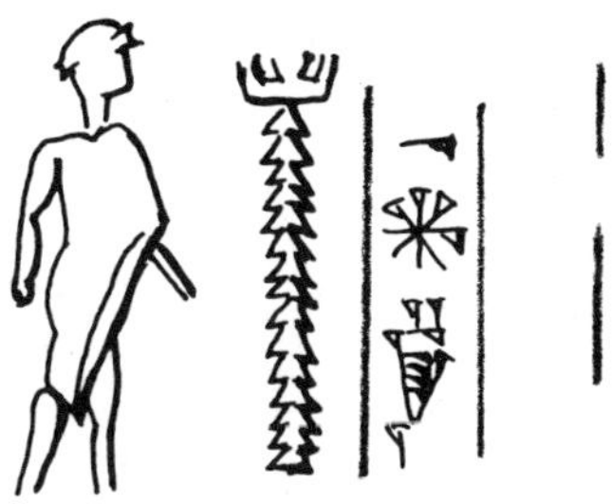

4 (290)

5 (292)

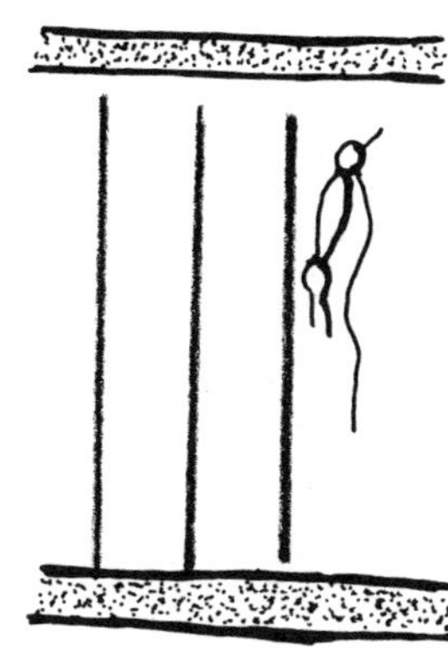

6 (289)

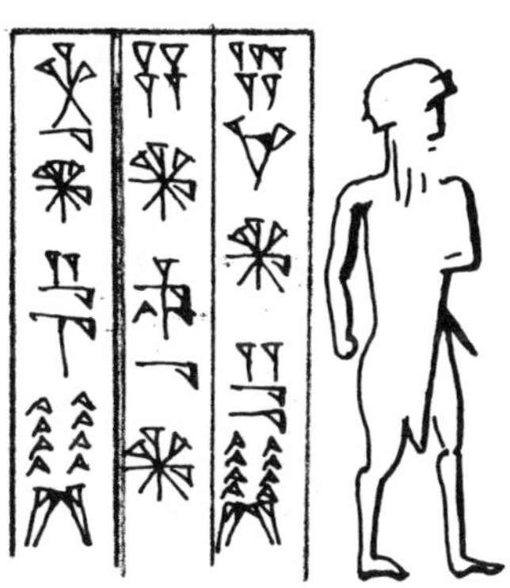

7 (288)

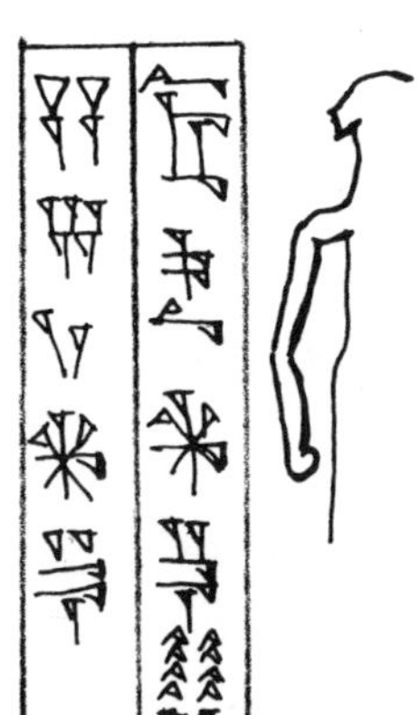

8 (291)

9 (293)

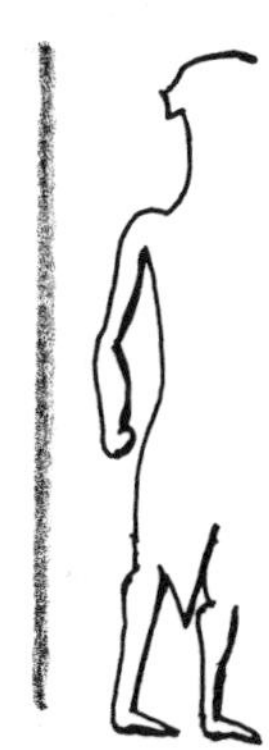

10 (294.a)

1 (319)

2 (303)

3 (304.a)

4 (309)

5 (316)

6 (320.a)

7 (297)

8 (308)

9 (315)

1 (298)

2 (299)

3 (304.b)

4 (318)

5 (322)

6 (295)

7 (326.a)

8 (326.b)

1 (158.d)

2 (189)

3 (189.a)

4 (189.f)

5 (189 g)

6 (197.b)

7 (198.j)

8 (221.f)

9 (247.c)

ILLUSTRATIONS

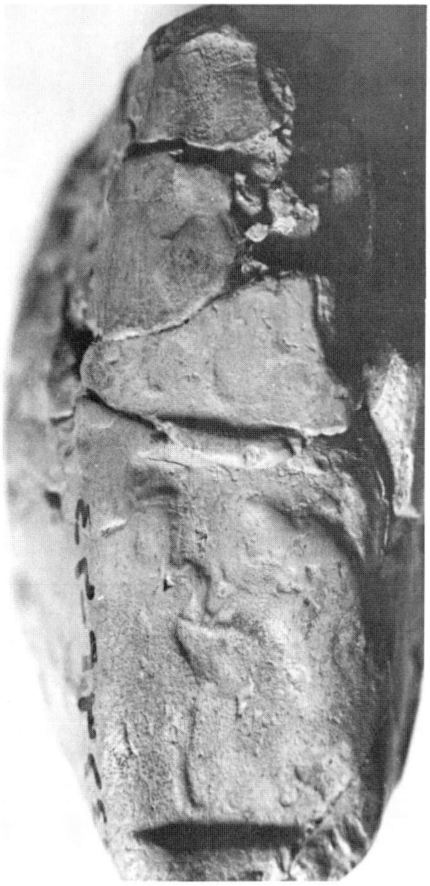

1.a

1.b

2.a

2.b

3.a

3.b and c

3.d

4.a

4.b

4.b

4.c

4.d

5

7

6

8.b

8.a

9

9

10.b

11.a

11.b

12.a

12.b

13.a and b

13.b

14

15.a

16.a, b and d

16.a and c

17.a and b

17.a

19

19

20.a and b

20.d

20.c

21.b

22

23

24

25.a

25.c

24

27.a

27.a

27.b

27.c

28.b

30

31.a

33.a

33.b

33.a

34.a

34.b

35

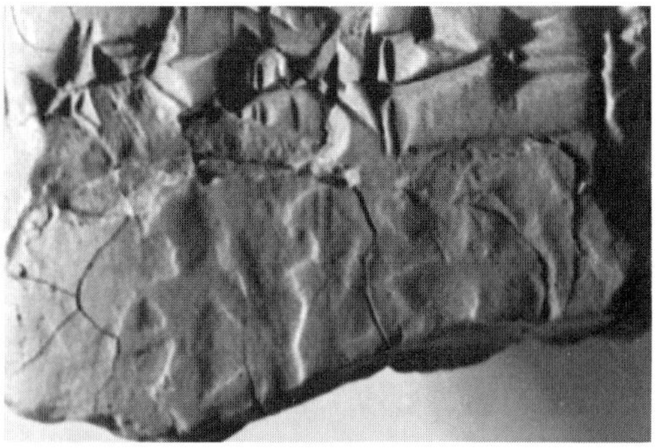

36.a

36.b

27

38

39.a

39.b

39.c

40

41.a

41.b

42.b

42.a

43

44

45

46.a

46.b

46.c

47.a

47.b

48.a

48.a

48.b

49.a

50.a

50.b

50.c

51.a

52

53.a

54.a, e and f

54.d

55.a and b

55.a

56.a

57.a and c

57.b

56.b

58

59.a

60.a

60.c

61

62.a

62.c

64

65.a

65.b

66

69.b

69.f

69.m

71

72

73

74

75

76.a

76.a

76.b and c

77.a and b

76.a

76.a

78

82

78

83

85

86

88

89

90

91

93.b

94

95

95

96

99

97

98

100

101

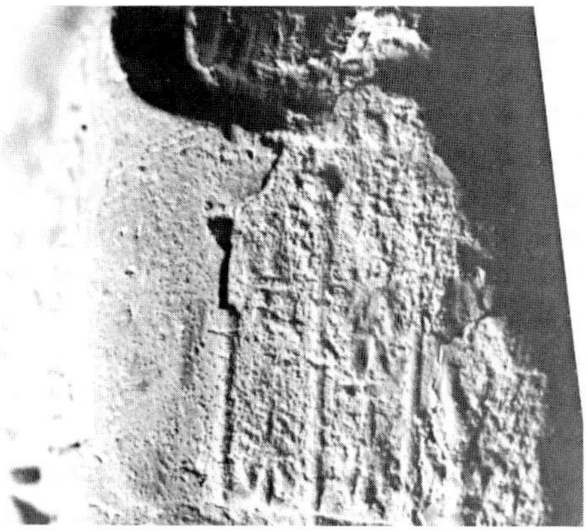

102.a

102.c

103

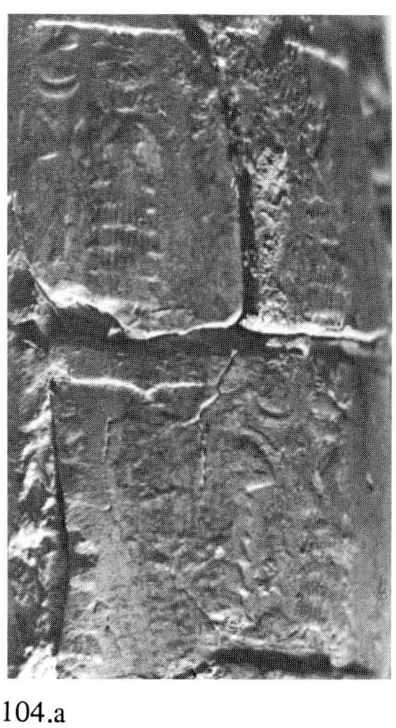

104.a

106

107

109

111

112

114.a

114.b

114.c

114.c

114.c

115

116

117

117

118

119

120

122

123.a

123.b

124

124

125

126

126

127

128.a

128.b.1

128.b.1

128.b.2

128.b.1

128.d

128.f

128.g

128.h

128.i

128.j

128.k

128.l

128.l

128.m

128.n

129.a

129.a

129.b

129.c

129.d

129.e

129.f

129.g

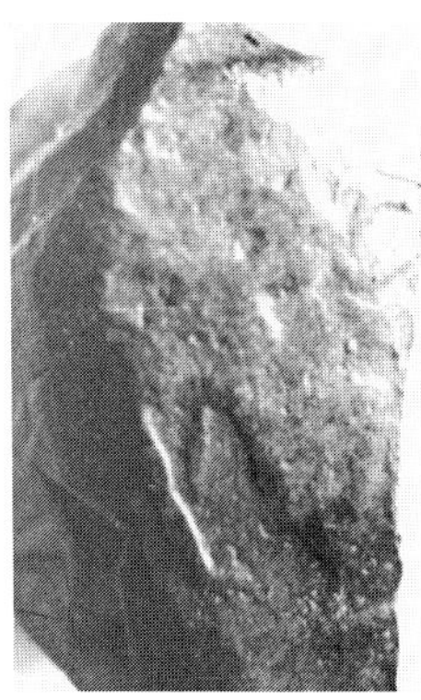

129.g

129.i

129.j

129.k

129.l

129.m

129.n

129.p

130.a

130.b

130.c

130.d

130.e

130.g

130.h

130.i

130j

130.k

130.m

131.a

131.a

131.b

131.c

131.b

131.e

131.f

131.f

131.g

131.g

131.h

131.i

131.i

131.j

131.k

131.l

131.m

131.n

131.n

131.o

131.p

131.q

131.s

131.v

131.t

132

133

134.a

134.b

135.b

134.b

136

137.a

137.b

137.b

138

138

139

140

141.a

141.b

142

142

143

144

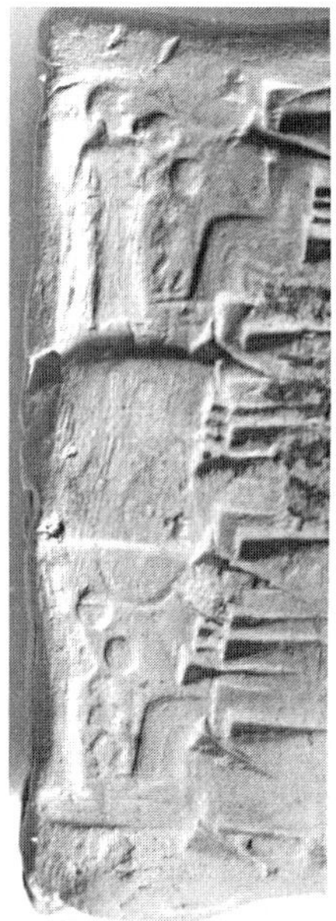

145

145

146

147

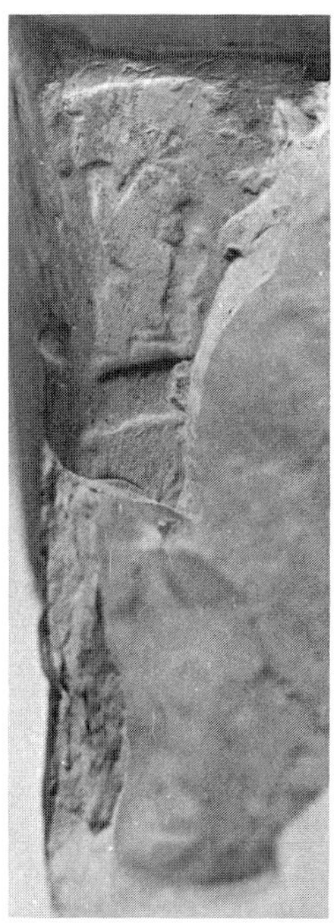

147

147

148

149

150

150

151

152.a

152.b

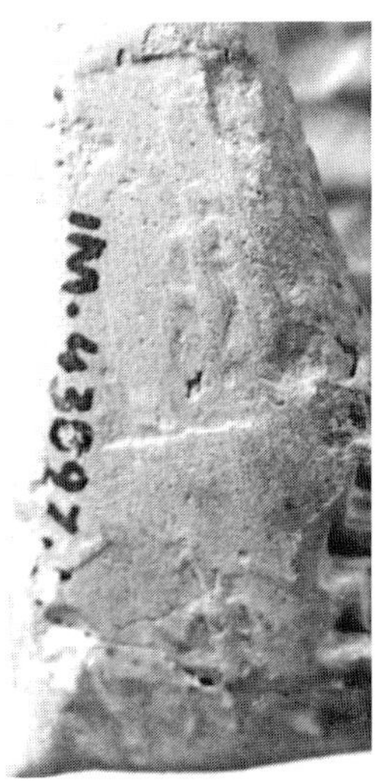

152.b

153

154.a

154.c

156

155

157.a

157.b

158.a and b

158.c

158.d

159.a

159.b

159.c

159.d

159.d

160.a

160.a

160.b

161

161

162

163

164

165.a

165.d

166.a

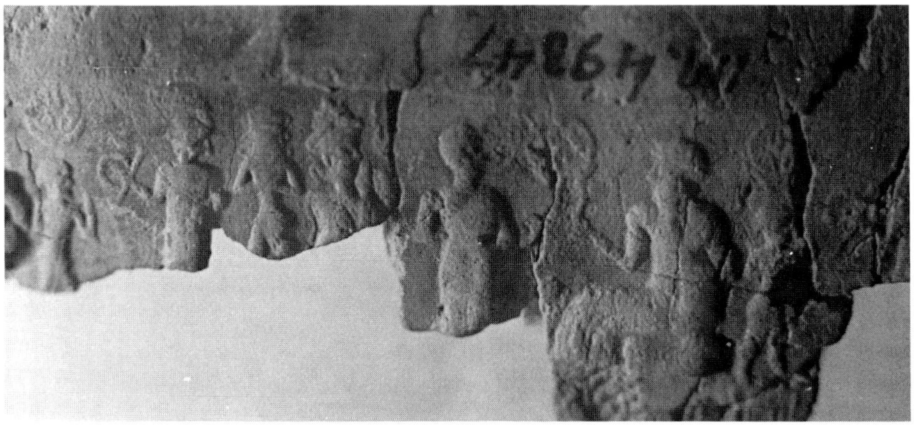

167

167

168

169

171

172

173

174

175

176

176

177

179.b

179.a

180

181

295

296

297

298

299

299

300

301

303

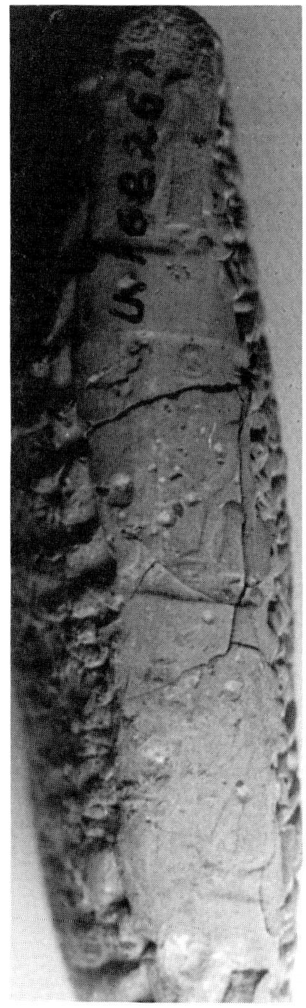

304

305

304.b

306

307

308

309

315

316

317

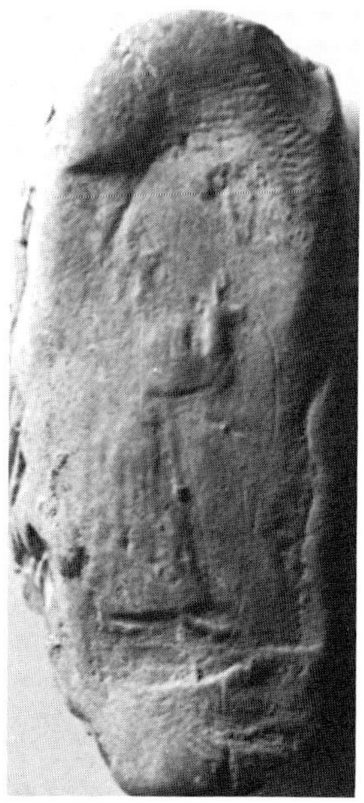

317

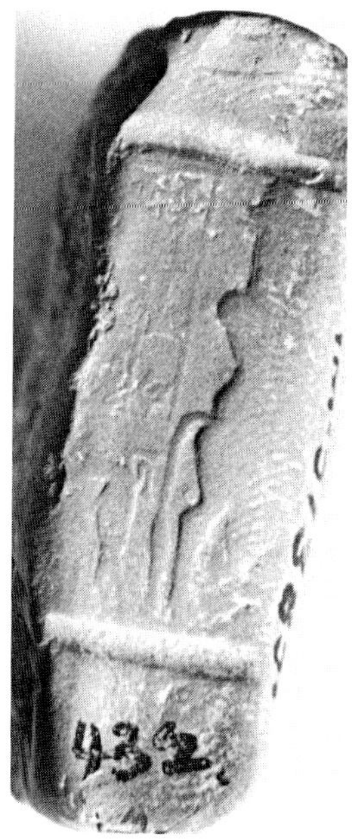

318

320

322

321

322

321

323

323

323

324

324

326

326